Library of Congress Cataloging in Publication Data

Main entry under title:

DC electronics.

"A Spectrum Book."
Reproduces the text of the ed. published: Benton
Harbor, Mich. : Heath Co., ©1978; programmed reviews
and unit objectives omitted.
Includes index.
1. Electric circuits—Direct current.   I. Heathkit/
Zenith Educational Systems (Group)
TK1111.D36 1983   621.319′12   82-23160
ISBN 0-13-198192-7
ISBN 0-13-198184-6 (pbk.)

This book is available at a special discount when ordered in bulk quantities. Contact Prentice-Hall, Inc., General Publishing Division, Special Sales, Englewood Cliffs, N.J. 07632.

This work is adapted from a larger work entitled *DC Electronics* © 1981 Heath Company. Revised Prentice-Hall edition © 1983.

A SPECTRUM BOOK

10   9   8   7   6   5   4   3   2   1

Printed in the United States of America

Manufacturing buyer Patrick Mahoney

ISBN 0-13-198192-7
ISBN 0-13-198184-6 {PBK.}

Prentice-Hall International, Inc., *London*
Prentice-Hall of Australia Pty. Limited, *Sydney*
Prentice-Hall of Canada Inc., *Toronto*
Prentice-Hall of India Private Limited, *New Delhi*
Prentice-Hall of Japan, Inc., *Tokyo*
Prentice-Hall of Southeast Asia Pte. Ltd., *Singapore*
Whitehall Books Limited, *Wellington, New Zealand*
Editora Prentice-Hall do Brasil Ltda., *Rio de Janeiro*

CONTENTS

Unit 1

CURRENT

INTRODUCTION

Electronics is that science which controls the behavior of *electrons* so that some useful function is performed. As this definition implies, the electron is vitally important to electronics. In fact, the word *electronics* is derived from the word *electron*. Electricity comes to our homes and offices by the movement of electrons through wires. Actually, electric current is nothing more than the movement of electrons. Obviously then, to understand electronics, we must first understand the nature of the electron. In this unit, we will see what the electron is, how it behaves, and how we can use it to perform useful jobs. We will also learn how to measure the flow of electrons.

COMPOSITION OF MATTER

Controlling the behavior of electrons is what electronics is all about. Therefore, an understanding of the electron is vitally important to an understanding of electronic fundamentals. *Electrons* are tiny particles which carry the energy to light our homes, cook our food, and do much of our work. To understand what an electron is, we must investigate the make-up of matter.

Matter is generally described as anything which has weight and occupies space. Thus, the earth and everything on it is classified as matter. Matter exists in three different states — solid, liquid, and gas. Examples of solid matter are gold, sand and wood. Some liquid examples are water, milk, and gasoline. Helium, hydrogen, and oxygen are examples of gas forms of matter.

Elements and Compounds

The basic building materials from which all matter is constructed are called *elements*. Hence, all matter is composed of elements. Some examples of elements are iron, carbon, hydrogen, and gold. Just over one hundred elements are presently known. Of these, only 92 occur in nature. These are called *natural elements*. Figure 1-1A lists the names of the 92 natural elements. In addition, there are about a dozen man-made elements. These are shown in Figure 1-1B.

As we look around us, it becomes obvious that there are many more types of matter than there are elements. For example, substances like salt, steel, water, and protein do not appear in the list of elements. The reason for this is that these substances are not elements. Instead, they are called *compounds*. A compound is a substance which is composed of two or more elements. Just as the letters of the alphabet can be arranged in various combinations to form millions of different words, the elements can be arranged in various combinations to form millions of different compounds. For example, water is a compound which is made up of the elements hydrogen and oxygen. On the other hand, sugar is composed of hydrogen, carbon, and oxygen while salt is composed of sodium and chlorine.

To better understand how the compound is related to its elements, let's investigate the structure of a compound with which we are all familiar — water. Suppose we divide a drop of water into two parts. Next, suppose we divide each part again and again. After a few dozen divisions, we have a drop so small that it can be seen only with a microscope. If we could

3

divide it even further into smaller and smaller particles, we would eventually get a particle so small that it could not be divided further and *still be water*. This smallest particle of water which still retains the characteristics of water is called a *molecule*. The water molecule can be broken into still smaller pieces but the pieces will not be water. Thus, if we break up the water molecule, we find that the pieces are the elements hydrogen and oxygen.

THE NATURAL ELEMENTS

Atomic Number	Name	Symbol	Atomic Number	Name	Symbol	Atomic Number	Name	Symbol
1	Hydrogen	H	32	Germanium	Ge	63	Europium	Eu
2	Helium	He	33	Arsenic	As	64	Gadolinium	Gd
3	Lithium	Li	34	Selenium	Se	65	Terbium	Tb
4	Beryllium	Be	35	Bromine	Br	66	Dysprosium	Dy
5	Boron	B	36	Krypton	Kr	67	Holmium	Ho
6	Carbon	C	37	Rubidium	Rb	68	Erbium	Er
7	Nitrogen	N	38	Strontium	Sr	69	Thulium	Tm
8	Oxygen	O	39	Yttrium	Y	70	Ytterbium	Yb
9	Fluorine	F	40	Zirconium	Zr	71	Lutetium	Lu
10	Neon	Ne	41	Niobium	Nb	72	Hafnium	Hf
11	Sodium	Na	42	Molybdenum	Mo	73	Tantalum	Ta
12	Magnesium	Mg	43	Technetium	Tc	74	Tungsten	W
13	Aluminum	Al	44	Ruthenium	Ru	75	Rhenium	Re
14	Silicon	Sl	45	Rhodium	Rh	76	Osmium	Os
15	Phosphorus	P	46	Palladium	Pd	77	Iridium	Ir
16	Sulfur	S	47	Silver	Ag	78	Platinum	Pt
17	Chlorine	Cl	48	Cadmium	Cd	79	Gold	Au
18	Argon	A	49	Indium	In	80	Mercury	Hg
19	Potassium	K	50	Tin	Sn	81	Thallium	Tl
20	Calcium	Ca	51	Antimony	Sb	82	Lead	Pb
21	Scandium	Sc	52	Tellurium	Te	83	Bismuth	Bi
22	Titanium	Ti	53	Iodine	I	84	Polonium	Po
23	Vanadium	V	54	Xenon	Xe	85	Astatine	At
24	Chromium	Cr	55	Cesium	Cs	86	Radon	Rn
25	Manganese	Mn	56	Barium	Ba	87	Francium	Fr
26	Iron	Fe	57	Lanthanum	La	88	Radium	Ra
27	Cobalt	Co	58	Cerium	Ce	89	Actinium	Ac
28	Nickel	Ni	59	Praseodymium	Pr	90	Thorium	Th
29	Copper	Cu	60	Neodymium	Nd	91	Protactinium	Pa
30	Zinc	Zn	61	Promethium	Pm	92	Uranium	U
31	Gallium	Ga	62	Samarium	Sm			

A

THE ARTIFICIAL ELEMENTS

Atomic Number	Name	Symbol	Atomic Number	Name	Symbol	Atomic Number	Name	Symbol
93	Neptunium	Np	97	Berkelium	Bk	101	Mendelevium	Mv
94	Plutonium	Pu	98	Californium	Cf	102	Nobelium	No
95	Americium	Am	99	Einsteinium	Es	103	Lawrencium	Lr
96	Curium	Cm	100	Fermium	Fm	104	Rutherfordium	Rf

B

Figure 1-1
Table of elements.

OXYGEN ATOM

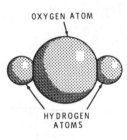

HYDROGEN ATOMS

Figure 1-2
The water molecule.

Atoms

The smallest particle to which an *element* can be reduced is called an *atom*. Molecules are made up of atoms which are bound together. The water molecule is shown in Figure 1-2 as three atoms. The two smaller atoms represent hydrogen while the large one represents oxygen. Therefore, a molecule of water consists of two atoms of hydrogen (H) and one atom of oxygen (O). This is the reason that the chemical formula for water is H_2O.

Electrons, Protons, and Neutrons

As small as the atom is, it can be broken up into even smaller particles. If we investigate the structure of the atom, we find that it is composed of three elementary particles. These particles are called *electrons, protons,* and *neutrons.* These are the three basic building blocks which make up all atoms and, therefore, all matter. Electrons, protons, and neutrons have very different characteristics. However, as far as is known, all electrons are exactly alike. By the same token, all protons are exactly alike. And finally, all neutrons are exactly alike.

Bohr Model of Atom

Figure 1-3 shows how electrons, protons, and neutrons are combined to form an atom. This particular one is a helium atom. Two protons and two neutrons are bunched together near the center of the atom. The center part of the atom which is composed of protons and neutrons is called the *nucleus.* Depending on the type of atom, the nucleus will contain from one to about 100 protons. Also, in all atoms except hydrogen, the nucleus contains neutrons. The neutrons and protons have approximately the same weight and size. The overall weight of the atom is determined primarily by the number of protons and neutrons in the nucleus.

Rotating around the nucleus are the electrons. Notice that the helium atom has two electrons. The electrons are extremely light and they travel at fantastic speeds. The atom can be compared to the solar system with the nucleus representing the sun and the electrons representing the planets. The electrons orbit the nucleus in much the same way that the planets orbit the sun.

It is interesting to note that no one has ever seen an atom because of its small size. Thus, any picture of the atom must be based on assumptions rather than actual observation. Figure 1-3 represents a very simple picture of the atom based on these assumptions. Today, much more complex models of the atom have been proposed. However, all these models have several things in common. They all assume that the basic structure is that of electrons orbiting about a nucleus which is composed largely of protons and neutrons. Thus, the model shown in Figure 1-3 is adequate for our purposes even though it may be somewhat simplified. This model of the atom is called the Bohr model after the man who proposed it.

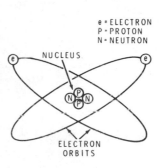

e = ELECTRON
P = PROTON
N = NEUTRON

NUCLEUS

ELECTRON ORBITS

Figure 1-3
Bohr model of the helium atom.

HYDROGEN ATOM
(one electron, one proton)

Difference Between Elements

Figure 1-4 shows the Bohr model of three different atoms. The first is hydrogen. It is the simplest atom of all. It consists of a single electron orbiting a nucleus which is composed of a single proton. This is the only atom which does not contain a neutron. Because of the simple structure of its atom, hydrogen is the lightest of all elements.

Also shown in Figure 1-4 is the Bohr model of the carbon atom. It contains 6 electrons orbiting a nucleus of 6 protons and 6 neutrons. Finally, the copper atom is shown. It contains 29 electrons and a nucleus composed of 29 protons and 35 neutrons. Although not shown in Figure 1-4, the most complex atom found in nature is the uranium atom. It consists of 92 electrons, 92 protons, and 146 neutrons. As you may have guessed, the difference between the various elements is that each is made up of atoms which contain a unique number of electrons, protons, and neutrons.

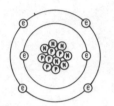

CARBON ATOM
(6 electrons, 6 protons, 6 neutrons)

Since there are only 92 natural elements, there are only 92 different types of atoms normally found in nature. The simplest is hydrogen with a single proton, the most complex is uranium with 92 protons.

The Balanced Atom

In the examples shown, you may have noticed that the number of electrons is always equal to the number of protons. This is normally true of any atom. When this is the case, the atom is said to be in its *normal*, *balanced*, or *neutral* state. As we will see later, this state can be upset by an external force. However, we normally think of the atom as containing equal numbers of electrons and protons.

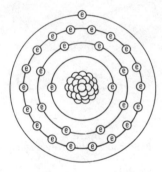

COPPER
(29 electrons, 29 protons, 35 neutrons)

Figure 1-4
The difference between atoms is
the number of electrons, protons,
and neutrons that they have.

6

ELECTROSTATICS

Electrostatics is the branch of physics dealing with electrical charges at rest, or static electricity. On the other hand, electronics deals largely with moving electrical charges. However, before we can fully understand the action of electrical charges in motion, we must first have some basic knowledge of their behavior at rest.

The Electrical Charge

We have examined the structure of the atom and discussed some of the characteristics of the electron, proton, and neutron. However, we have not yet discussed the most important characteristic of these particles. This characteristic is their *electrical charge*. An electrical charge is a property associated with the electron and the proton. It is this electrical charge which makes the electron useful in electrical and electronic work.

The electrical charge is difficult to visualize because it is not a thing, like a molecule or an atom. Rather, it is a property which electrons and protons have that causes these particles to behave in certain predictable ways.

There are two distinct types of electrical charges. Because these two types of charges have opposite characteristics, they have been given the names *positive* and *negative*. The electrical charge associated with the *electron* has been arbitrarily given the name *negative*. On the other hand, the electrical charge associated with the *proton* is considered *positive*. The neutron has no electrical charge at all. It is electrically neutral and, therefore, plays no known role in electricity.

The electron revolves around the nucleus of the atom in much the same way that the earth orbits the sun. Let's compare this action to that of a ball which is attached to the end of a string and twirled in a circle. If the string breaks, the ball will fly off in a straight line. Thus, it is the restraining action of the string which holds the path of the ball to a circle. In the case of the earth rotating around the sun, it is the gravitational attraction of the sun which prevents the earth from flying off into space. The gravitational attraction of the sun exactly balances the centrifugal force of each planet. Thus, the planets travel in more or less circular paths around the sun.

7

The electron orbits around the nucleus at a fantastic speed. What force keeps the electron from flying off into space? It is not gravity because the gravitational force exerted by the nucleus is much too weak. Instead, the force at work here is caused by the charge on the electron in orbit and the charge on the proton in the nucleus. The negative charge of the electron is *attracted* by the positive charge of the proton. We call this force of attraction an *electrostatic force*. To explain this force, science has adopted the concept of an *electrostatic field*. Every charged particle is assumed to be surrounded by an electrostatic field which extends for a distance outside the particle itself. It is the interaction of these fields which causes the electron and proton to attract each other.

Figure 1-5A shows a diagram of a proton. The plus sign represents the positive electrical charge. The arrows which extend outward represent the lines of force which make up the electrostatic field. Notice that the lines are arbitrarily assumed to extend outward away from the positive charge. Compare this to the electron shown in Figure 1-5B. The minus sign represents the negative charge while the arrows which point inward represent the lines of the electric field. Now let's see how these two fields interact with one another.

Law of Electrical Charges

There is a basic law of nature which describes the action of electrical charges. It is called Coulomb's Law after Charles A. de Coulomb who discovered this relationship.

Quite simply, Coulomb's Law states that:
1. Like charges repel
2. Unlike charges attract

A   PROTON

B   ELECTRON

Figure 1-5

Fields associated with protons and electrons.

Because like charges repel, two electrons repel each other as do two protons. Figure 1-6A illustrates how the lines of force interact between two electrons. The directions of the line of force are such that the two fields cannot interconnect. The net effect is that the electrons attempt to move apart. That is, they repel each other. Figure 1-6B illustrates that the same is true of two protons. In Figure 1-6C, an electron and a proton are shown. Here, the two fields do interconnect. As a result, the two charges attract and tend to move together.

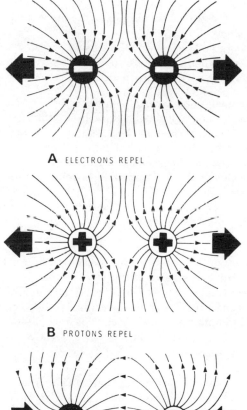

A ELECTRONS REPEL

B PROTONS REPEL

Figure 1-6
Action of like and unlike charges.

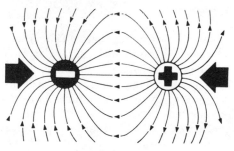

C ELECTRON AND PROTON ATTRACT

These examples show only individual charged particles. However, Coulomb's Law holds true for concentrations of charges as well. In fact, it holds true for any two charged bodies. An important part of Coulomb's Law is an equation which allows us to determine the force of attraction or repulsion between charged bodies. The equation states that:

$$F = \frac{q_1 \times q_2}{d^2}$$

Where
F = the force of attraction between unlike charges or the force of repulsion between like charges
q_1 = the charge on one body
q_2 = the charge on the second body
d^2 = the square of the distance between the two bodies.

While we need not work actual problems to determine the force between charges, we can see some interesting relationships by examining the equation. If we experiment with the equation by substituting some simple arbitrary numbers for q_1, q_2, and d^2, we can determine how the force changes as the quantities change. For example, if the value of either charge doubles, the force also doubles. If both charges double, then the force increases by a factor of four. On the other hand, increasing the distance between charges decreases the force. If the distance between charges is doubled, the force is reduced to one fourth its former value.

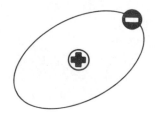

Figure 1-7

Hydrogen atom.

The magnitude of the negative charge on the electron is exactly equal to the magnitude of the positive charge on the proton. Figure 1-7 is a diagram of a hydrogen atom consisting of one electron in orbit around one proton. Notice that the negative charge of the electron is exactly offset by the positive charge of the proton. Thus, the atom as a whole has no charge at all. That is, overall, this atom has neither a negative nor a positive charge. It is electrically neutral.

Atoms which are electrically neutral have no net charge. Therefore, they neither attract nor repel each other. By the same token they are neither attracted nor repelled by charged particles such as electrons and protons. We have seen that atoms normally contain the same number of electrons (negative charges) as protons (positive charges). And, since the neutrons add no charge, all atoms are normally neutral as far as their electrical charges are concerned. However, this *normal* condition can be easily upset by external forces.

The Ion

Atoms are affected by many outside forces such as heat, light, electrostatic fields, chemical reactions and magnetic fields. Quite often the balanced state of the atom is upset by one or more of these forces. As a result, an atom can lose or gain an electron. When this happens, the number of negative charges is no longer exactly offset by the number of positive charges. Thus, the atom ends up with a net charge. An atom which is no longer in its neutral state is called an ion. The process of changing an atom to an ion is called *ionization*.

There are both negative and positive ions. Figure 1-8 compares a neutral atom of carbon with negative and positive ions of carbon. Figure 1-8A shows the balanced or neutral atom. Notice that the six negative charges (electrons) are exactly offset by the six positive charges (protons). The neutrons are ignored in this example since they contribute nothing to the electrical charge.

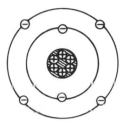

A NEUTRAL ATOM

Figure 1-8B shows the condition which exists when the carbon atom loses an electron. There are many forces in nature which can dislodge an electron and cause it to wander away from the atom. We will discuss this in more detail later. Notice that the carbon atom now has one more proton than electrons. Thus, there is one positive charge which is not cancelled by a corresponding negative charge. Therefore, the atom has a net positive charge. We call this a positive ion.

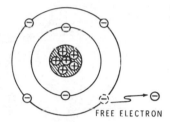

FREE ELECTRON

B POSITIVE ION CAUSED
BY LOSING ONE ELECTRON

Figure 1-8C shows a carbon atom which has picked up a stray electron. In this case, there is one negative charge which is not offset by a corresponding positive charge. Hence, the atom has a net negative charge. This is called a negative ion.

The ion still has all the basic characteristics of carbon because the nucleus of the atom has not been disturbed. Therefore, an atom can give up or pick up electrons without changing its basic characteristics.

Changing atoms to ions is an easy thing to do and everything you see around you contains ions as well as atoms. The material around you also contains a large number of free or stray electrons. These are electrons which have escaped from atoms leaving behind a positive ion. As we will see later, the electrical characteristics of different types of material are determined largely by the number of free electrons and ions within the material.

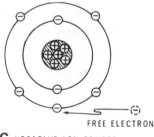

FREE ELECTRON

C NEGATIVE ION CAUSED
BY PICKING UP A
STRAY ELECTRON

Figure 1-8

Carbon atom and ions.

Action of Electrostatic Charges

At one time or another we have all seen or felt the effects of electrostatic charges. A most spectacular effect which we have all seen is lightning. Less spectacular examples are often witnessed when we remove clothes from a dryer, comb our hair, or touch a metal object after scuffing our feet on a rug. In each of these cases, two different bodies receive opposite electrical charges. This is caused by one of the bodies giving up a large number of electrons to the other. The body which gives up the electrons becomes positively charged while the body receiving the electrons becomes negatively charged.

When we comb our hair vigorously with a hard rubber comb, our hair gives up electrons to the comb. Thus, the comb becomes negatively charged while our hair becomes positively charged. That is, the comb collects a large number of free electrons from our hair. This is an example of *charging by friction.*

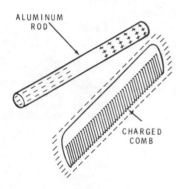

Figure 1-9
Charging by induction.

There are several other ways in which an object can be charged. For example, the charge on the comb can be partially transferred to another body simply by touching the comb to the uncharged body. When the charged comb comes into contact with the uncharged object, many of the excess electrons leave the comb and collect on the other object. If we now remove the comb, the object will have a charge of its own. This is called *charging by contact.*

Another method of charging is called *charging by induction.* This method takes advantage of the electrostatic field which exists in the space surrounding a charged body. This allows us to charge an object without actually touching it with a charged body. Figure 1-9 shows the negatively charged comb placed close to an aluminum rod. The excess electrons in the comb repel the free electrons in the rod. Consequently, the free electrons gather at the end of the rod away from the charged comb. This causes that end of the rod to acquire a negative charge. The opposite end acquires a positive charge because of the deficiency of electrons. If we now touch the negative end of the rod with a neutral body, some of the electrons leave the rod and enter the neutral body. This leaves the rod with a net positive charge. Thus, we have induced a positive charge into the rod without touching it with a charged body.

Now, let's see how electrical charges can be neutralized. When a glass rod is rubbed with a silk cloth, the glass gives up electrons to the silk. Therefore, the glass becomes positively charged while the silk becomes negatively charged. This is shown in Figure 1-10A. However, if the rod is now brought back into contact with the cloth, the negative electrons in the silk are attracted by the positive charge in the glass. The force of attraction pulls the electrons back out of the silk so that the charge is neutralized as shown in Figure 1-10B. Thus, if two objects having equal but opposite charges are brought into contact, electrons flow from the negatively charged object into the positively charged object. The flow of electrons continues until both charges have been neutralized.

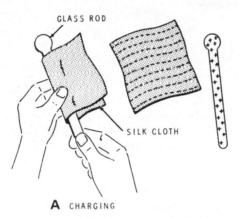

GLASS ROD

SILK CLOTH

A CHARGING

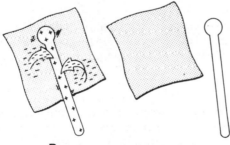

B NEUTRALIZING THE CHARGE

Figure 1-10
Charging and discharging a
glass rod.

CURRENT FLOW

In electronics, current is defined as the flow of electrical charge from one point to another. We have already seen some examples of this. We saw that when a negatively charged body is touched to a positively charged body, electrons flow from the negative object to the positive object. Since electrons carry a negative charge, this is an example of electrical charges flowing. Before an electron can flow from one point to another, it must first be freed from the atom. Therefore, let's take a closer look at the mechanism by which electrons are dislodged from the atom.

Freeing Electrons

We have seen that electrons revolve around the nucleus at very high speeds. Two forces hold the electron in a precarious balance. The centrifugal force of the electron is exactly offset by the attraction of the nucleus. This balanced condition can be upset very easily so that the electron is dislodged.

Not all electrons can be freed from the atom with the same ease. Some are dislodged more easily than others. To see why, we must discuss the concept of *orbital shells*. It is believed that the orbits of the electrons in an atom fall in a certain pattern. For example, in all atoms which have two or more electrons, two of the electrons orbit relatively close to the nucleus. The area in which these electrons rotate is called a *shell*. The shell closest to the nucleus contains two electrons. This area can support only two electrons and all other electrons must orbit in shells further from the nucleus.

A second shell somewhat further from the nucleus can contain up to eight electrons. Also there is a third shell which can contain up to 18 electrons and a fourth shell which can hold up to 32 electrons. The first four shells are illustrated in Figure 1-11. Although not shown, there are also additional shells in the heavier atoms.

Of particular importance to electronics is the outer electron shell of the atom. Hydrogen and helium atoms have one and two electrons respectively. In this case, the outer shell is the first (and only) shell. With atoms which have three to ten electrons, the outer shell is the second shell. Regardless of which shell it happens to be, the outer shell is called the *valence shell*. Electrons in this shell are called *valence electrons*.

Electrons are arranged in such a way that the valence shell never has more than eight electrons. This may be confusing since we have seen that the third shell can contain up to 18 electrons. An example shows why both

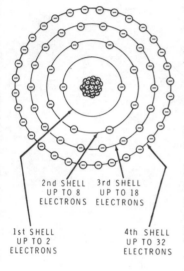

2nd SHELL
UP TO 8
ELECTRONS

3rd SHELL
UP TO 18
ELECTRONS

1st SHELL
UP TO 2
ELECTRONS

4th SHELL
UP TO 32
ELECTRONS

Figure 1-11
Arrangement of orbital
shells in atom.

statements are true. An atom of argon contains 18 electrons — 2 in the first shell, 8 in the second shell, and 8 in the third shell. It might seem that the next heavier element, potassium, would have 9 electrons in its third shell. However, this would violate the valence rule stated above. Actually, what happens is that the extra electron is placed in a fourth shell. Thus, the 19 electrons are distributed in this manner — 2 in the first shell, 8 in the second shell, 8 in the third shell, and 1 in the fourth shell. Notice that the outer or valence shell becomes the fourth shell rather than the third. Once the fourth shell is established as the valence shell, the third shell can fill to its full capacity of 18 electrons.

The valence electrons are extremely important in electronics. These are the electrons which can be easily freed to perform useful functions. To see why the valence electrons are more easily freed, let's consider the structure of an atom of copper. Figure 1-12A shows how the electrons are distributed in the various shells. Notice that the valence shell contains only one electron. This electron is further from the nucleus than any of the other electrons. From Coulomb's Law we know that the force of attraction between charged particles decreases as the distance increases. Therefore the valence electrons experience less attraction from the nucleus. For this reason, these electrons can be easily dislodged from the atom.

Since we are concerned primarily with the valence electrons, we need not show the inner electrons. Instead, we can show the atom in the simplified form shown in Figure 1-12B. Figures 1-12C and D use this simplified form to illustrate one way in which a valence electron can be freed. Here two copper atoms are shown as they might appear in a copper wire. Each valence electron is held in orbit by the attraction of the nucleus. However, the force of attraction is quite weak because the orbits are so far from the nucleus. If these two atoms are close together, the valence shells may be closer together than either electron is to its nucleus. At certain points in their orbits the two electrons may come very close together. When this happens the force of repulsion between the two electrons is stronger than the force of attraction exerted by the nucleus. Thus, one or both of the electrons may be forced out of orbit to wander as a free electron. Notice that when the electron leaves, the atom becomes a positive ion.

As the free electron wanders around through the atomic structure, it may be eventually captured by another positive ion. Or, it may come close enough to other valence electrons to force them from orbit. The point is that events like these occur frequently in many types of material. Thus, in a piece of copper wire containing billions and billions of atoms, there are bound to be billions of free electrons wandering around the atomic structure.

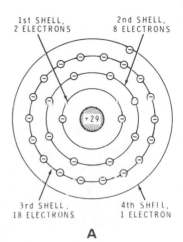

1st SHELL,
2 ELECTRONS

2nd SHELL,
8 ELECTRONS

3rd SHELL,
18 ELECTRONS

4th SHELL,
1 ELECTRON

A

B

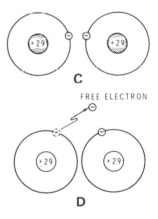

C

FREE ELECTRON

D

Figure 1-12
Freeing an electron from a copper atom.

15

Conductors and Insulators

The importance of the valence electrons cannot be emphasized too strongly. Both the electrical and the chemical characterisitics of the elements depend on the action of the valence electrons. An element's electrical and chemical stability are determined largely by the number of electrons in the valence shell. We have seen that the valence shell can contain up to eight electrons. Those elements which have valence shells that are filled or nearly filled tend to be stable. For example, the elements neon, argon, krypton, xenon, and radon have 8 electrons in their valence shell. Thus, the valence shell is completely filled. As a result, these elements are so stable that they resist any sort of chemical activity. They will not even combine with other elements to form compounds. Also, atoms of these elements are very reluctant to give up electrons. All these elements have similar characteristics in that they are all inert gases.

Elements which have their valence shells *almost* filled tend to be stable also, although they are not as stable as those whose valence shell is completely filled. These elements will strive to fill their valence shell by capturing free electrons. Consequently, elements of this type have very few free electrons wandering around through the atomic structure. Substances which have very few free electrons are called *insulators*. In addition to certain elements which act as insulators, there are many compounds which have few free electrons. Thus, they act as insulators also. By opposing the production of free electrons, these substances resist certain electrical actions. Insulators are important in electrical and electronics work for this reason. The plastic material on electrical wires is an insulator which protects us from electrical shock.

Elements in which the valence shell is almost empty have the opposite characteristics. Those which have only one or two electrons tend to give up these electrons very easily. For example, copper, silver, and gold each have one valence electron. In these elements, the valence electrons are very easily dislodged. Consequently, a bar of any one of these elements will have a very large number of free electrons. Substances which have a large number of free electrons are called *conductors*. In addition to silver, copper, and gold, some other good conductors are iron, nickel, and aluminum. Notice that all of these elements are metals. Most metals are good conductors. Conductors are important because they are used to carry electrical current from one place to another.

In some elements, the valence shell is half filled. That is, there are four valence electrons. Two examples of elements of this type are silicon and germanium. We call these elements *semiconductors* because they are neither good conductors nor good insulators. Semiconductors are important in electronics because transistors and integrated circuits are composed of these elements. However, in this course, we will be concerned primarily with conductors and insulators.

The Battery

Current flow is the movement of free electrons from one place to another. Thus, to have current flow we must first have free electrons. We have seen how valence electrons can be dislodged from atoms to form free electrons and positive ions. This can be done by very simple means such as combing our hair or rubbing a glass rod with a silk cloth. However, to perform a useful function, we must free very large numbers of electrons and concentrate them in one area. This requires more sophisticated techniques. One device for doing this is the ordinary battery. There are many different types of batteries. Figure 1-13 shows two familiar examples. These are the dry cell (flash-light battery) and the wet cell (automobile battery).

While these two types of batteries are quite different in construction, they do have several points in common. Both have two terminals or poles to which an electrical circuit can be connected. Also, both employ a chemical reaction which produces an excess of electrons at one terminal and a deficiency of electrons at the other. The terminal at which the electrons congregate is called the negative terminal. It is indicated by the minus sign in Figure 1-13. The other terminal is indicated by a plus sign and has a deficiency of electrons. Now let's see how the battery affects the free electrons in a conductor.

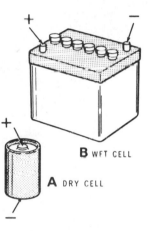

Figure 1-13
Types of batteries.

Random Drift and Directed Drift

A conductor is a substance which has a large number of free electrons. In a conductor, the free electrons do not stand still. Instead they drift about in a random motion. Figure 1-14A represents a small section of a conductor containing many free electrons. At any instant, the free electrons are drifting at random in all directions. This is referred to as *random drift*. This type of drift occurs in all conductors but it has little practical use. To do useful work, the free electrons must be forced to drift in the same direction rather than at random.

We can influence the drift of electrons so that all or most electrons move in the same direction through the conductor. This can be done by placing electrical charges at opposite ends of the conductor. Figure 1-14B shows

17

a negative charge placed at one end of the conductor while a positive charge is placed at the other. The negative charge repels the free electrons while the positive charge attracts them. As a result all of the free electrons move or drift in the same general direction. The direction is from the negative charge to the positive charge.

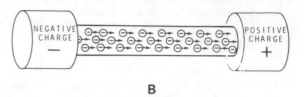

Figure 1-14

Comparison of random and directed drift.

A   **B**

Here, the application of the electrical charges at the ends of the conductor has changed random drift to directed drift. This directed drift of free electrons is called *current flow*. We say that an *electric current* is flowing through the conductor. If the electrical charges shown in Figure 1-14B are isolated from one another, the flow of electrons will quickly cancel both charges and only a momentary current will flow. However, if the two electrical charges are caused by a battery, the chemical action of the battery can maintain the two charges for some time. Therefore, a battery can maintain a continuous current through a conductor for a long period.

A copper wire is a good example of a conductor. Figure 1-15 shows a length of copper wire connected from one terminal to the other of a battery. A heavy current will flow from the negative terminal of the battery to the positive terminal. Recall that the negative terminal is a source of free electrons. An electron at this point is repelled by the negative charge and is attracted by the positive charge at the opposite terminal. Thus, the electrons flow through the wire as shown. When they enter the positive terminal of the battery, they are captured by positive ions. The chemical reaction of the battery is constantly releasing new free electrons and positive ions to make up for the ones lost by recombination.

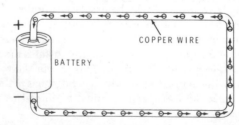

Figure 1-15

Current flows from the negative terminal to the positive terminal of the battery.

It should be pointed out that in practice, we never connect a conductor directly across the terminals of the battery as shown in Figure 1-15. The heavy current would quickly exhaust the battery. This is an example of a "short circuit" and is normally avoided at all cost. This example is shown here merely to illustrate the concept of current flow.

THE ELECTRIC CIRCUIT

In its simplest form, an electric circuit consists of a power source, a load, and conductors for connecting the power source to the load. Often the power source is a battery. The purpose of the power source is to provide the force necessary to direct the flow of electrons. As you will see in the next unit, this force is called *voltage*. Power sources produce voltage by creating a positive charge at one terminal and a negative charge at the other.

The *load* is generally some kind of electrical device which performs a useful function. It might be a lamp which produces light, a motor which produces physical motion, a horn which produces sound, or a heating element which produces heat. Regardless of the type of load used, the load performs its useful function only when electric current flows through it.

The third part of the circuit is the conductors which connect the power source to the load. They provide a path for current flow. The conductor may be a length of copper wire, a strip of aluminum, the metal frame of an automobile, etc.

Figure 1-16 shows a pictorial representation of an electric circuit consisting of a battery, a lamp, and connecting copper wires. The battery produces the force (voltage) necessary to cause the directed flow of electrons. The force developed by the battery causes the free electrons in the conductor to flow through the lamp in the direction shown. The free electrons are repelled by the negative charge and are attracted by the positive charge. Thus, the electrons flow from negative to positive. The negative and positive charges in the battery are constantly being replenished by the chemical action of the battery. Therefore, the battery can maintain a current flow for a long period of time. As the electrons flow through the lamp, they heat up the wire within the lamp. As the wire becomes hotter, the lamp emits light. The lamp will glow as long as a fairly strong current is maintained.

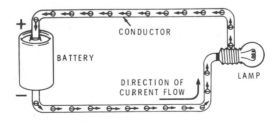

Figure 1-16
Simple electric circuit.

19

We know from our experiences with flashlights that a battery cannot maintain a constant current flow forever. As the battery is used, the chemical reaction within the battery slows down. Over a period of time, the force provided by the battery becomes weaker and less current is provided. As a result the lamp emits less light. It becomes dimmer and dimmer and eventually it emits no light at all. At this time the battery is said to be dead, burned out, or run down. In this condition the battery cannot produce the force necessary to push enough electrons through the lamp to cause the lamp to glow.

The circuit in Figure 1-16 can be made much more practical by adding one additional component. This component is a switch which provides a simple method of turning the lamp on and off.

Figure 1-17 shows the circuit after the switch has been added. For simplicity, a "knife" switch is shown. It consists of two metal contacts to which conductors may be connected, a metal arm which can be opened and closed, and a base. Current cannot flow through the base of the switch because an insulator material is used. Current can flow only through the arm and then only if the arm is closed.

In Figure 1-17A, the switch is shown closed. With the switch closed, there is a path for current flow from the negative terminal of the battery through the switch and lamp to the positive terminal. The lamp lights because current flows through it. When the switch is opened as shown in Figure 1-17B, the path for current flow is broken. Thus, the lamp does not glow because there is no current flowing through it.

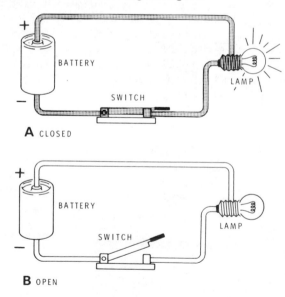

Figure 1-17

Circuit with switch.

While simple circuits can be drawn as shown in Figures 1-16 and 1-17, it would be very difficult to draw complex circuits in this manner. For this reason, the *schematic diagram* was developed. A schematic diagram is a drawing in which symbols are used to represent circuit components. Thus, the first step to understanding the schematic diagram is to learn the symbols for the various components used. Figure 1-18 compares the schematic symbol with the pictorial representation of the circuit components we have used up to this point. The conductor is represented by a single line in the schematic. Also, the picture of the battery is replaced by a series of long and short lines. The long line represents the positive terminal while the short line represents the negative terminal. The same symbol can be used regardless of the type of battery. The symbols for the lamp and switch are also shown.

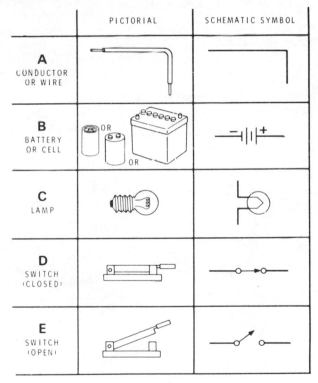

	PICTORIAL	SCHEMATIC SYMBOL
A CONDUCTOR OR WIRE		
B BATTERY OR CELL		
C LAMP		
D SWITCH (CLOSED)		
E SWITCH (OPEN)		

Figure 1-18
Pictorial representations compared
with the schematic symbols.

Figure 1-19 shows several of the symbols combined to form a schematic diagram. Figure 1-19A is the schematic diagram for the pictorial drawing shown earlier in Figure 1-17A. Also, Figure 1-19B is a schematic diagram of the pictorial shown in Figure 1-17B.

The circuit shown in Figure 1-19 is the schematic diagram of a flashlight. It is also the diagram for the headlight system in an automobile. In fact it can represent any system which contains a battery, a lamp, and a switch. If the lamp is replaced with a motor, the circuit becomes that of the starter system in a car. In this case, the switch is operated by the ignition key. Other circuits which operate in a similar manner are the doorbell and the automobile horn. In the first case, the bell is the load while the switch is operated by a push button at the door. In the second case, the horn is the load while the switch is located on the steering wheel.

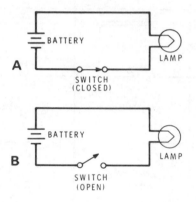

Figure 1-19
Schematic diagram of a simple circuit.

MEASURING CURRENT

Current is the flow of electrons from a negative to a positive charge. To measure current flow, we must measure the number of electrons flowing past a point in a specific length of time. Before we discuss how current is measured, we must first define the unit of electrical charge and the unit of current.

The Coulomb

We have seen that the charge on an object is determined by the number of electrons which the object loses or gains. If the object loses electrons, the charge is positive. However, an object which gains electrons has a negative charge. The unit of electrical charge is called the *coulomb*. The coulomb is equal to the charge of 6.25×10^{18} electrons. For those who are not used to expressing numbers in this way, the number is:

$$6,250,000,000,000,000,000.$$

An object which has gained 6.25×10^{18} electrons has a negative charge of one coulomb. On the other hand, an object which has given up 6.25×10^{18} electrons has a positive charge of one coulomb.

Powers of Ten and Scientific Notation

A word about powers of ten and scientific notation may be helpful at this point. The number 6,250,000,000,000,000,000 can be expressed as 6.25×10^{18}. This number is read "six point two five times ten to the eighteenth power." The expression "ten to the eighteenth power" means that the decimal place in 6.25 should be moved 18 places to the right in order to convert to the proper number. The theory is that it is easier to write and remember 6.25×10^{18} than it is to write and remember 6,250,000,000,000,000,000. This shorthand method of expressing numbers is known as powers of ten or scientific notation. It is often used in electronics to express very large and very small numbers. Very small numbers are expressed by using negative powers of ten. For example, 3.2×10^{-8} is scientific notation for the number 0.000000032. Here, "ten to the minus eighth power" means "move the decimal place in 3.2 eight places to the left." To be sure you have the idea, let's look at some examples of both positive and negative powers of ten:

Positive Powers of Ten

$7.9 \times 10^4 = 79,000$
$9.1 \times 10^8 = 910,000,000$
$1.0 \times 10^{12} = 1,000,000,000,000$

23

Negative Powers of Ten

$7.9 \times 10^{-4} = 0.00079$
$9.1 \times 10^{-8} = 0.000\ 000\ 091$
$1.0 \times 10^{-12} = 0.000\ 000\ 000\ 001$

Study these examples until you get the idea of this system of writing numbers. If you feel you need additional explanation, read Appendix A at the end of this unit. It is a programmed instruction sequence designed to teach powers of ten and scientific notation in much greater detail.

The Ampere

The unit of current is the *ampere*. The ampere is the rate at which electrons move past a given point. As mentioned above, 1 coulomb is equal to 6.25×10^{18} electrons. An ampere is equal to 1 *coulomb per second*. That is, if 1 coulomb (6.25×10^{18} electrons) flows past a given point in 1 second then the current is equal to 1 ampere. Coulombs indicate numbers of electrons; amperes indicate the rate of electron flow or coulombs per second.

When 6.25×10^{18} electrons flow through a wire each second, the current flow is 1 ampere. If twice this number of electrons flows each second, the current is 2 amperes. This relationship is expressed by the equation:

$$\text{amperes} = \frac{\text{Coulombs}}{\text{seconds}}$$

If 10 coulombs flow past a point in two seconds, then the current flow is 5 amperes.

The name ampere is often shortened to *amp* and is abbreviated A. Many times the ampere is too large a unit. In these cases metric prefixes are used to denote smaller units. The milliampere (mA) is one thousandth (.001) of an ampere. The microampere (μA) is one millionth (.000 001) of an ampere. In other words, there are 1000 milliamperes or 1,000,000 microamperes in an ampere.

We change from amperes to milliamperes by multiplying by 10^3. Thus, 1.7 amperes is equal to 1.7×10^3 milliamperes. Also, we change from amperes to microamperes by multiplying by 10^6. Therefore, 1.7 amperes is equal to 1.7×10^6 microamperes.

For those who need it, a more detailed explanation of metric prefixes is given in Appendix A.

The Ammeter

A device for measuring current flow is the *ammeter*. The name *ammeter* is a shortened form of the name ampere meter. Figure 1-20 shows a diagram of an ammeter. It has a pointer which moves in front of a calibrated scale. In this figure, the scale is calibrated from 0 to 10 amperes. The movement of the pointer is proportional to the amount of current flowing through the meter. Therefore, an accurate indication of the amount of current flowing in a circuit is obtained by reading the pointer against the scale. This meter is presently displaying a reading of just over 6 amperes.

Figure 1-21A shows a circuit in which an unknown amount of current is flowing. We can measure this current by inserting an ammeter into the circuit as shown in Figure 1-21B. Notice that the schematic symbol for the ammeter is a circle with the letter A. Before the ammeter can measure current, it must be placed in the circuit in such a way that the current we wish to measure actually flows through the meter. We say that the ammeter is connected in *series* with the circuit elements. Incidentally, a circuit like the one shown in Figure 1-21B is called a *series* circuit. A series circuit is one in which the same current flows through all the elements in one continuous loop.

The maximum current that an ammeter can safely measure is indicated by the highest number on the scale. The highest current that the ammeter in Figure 1-20 can safely measure is 10 amperes. This is called its full scale reading. Many current meters are much more sensitive. Some have a full scale reading of 1 milliampere. Others provide a full scale reading with only 50 microamperes flowing through them.

Ammeters are delicate instruments and can be destroyed if the current applied greatly exceeds the full scale reading of the meter. For this reason, we must exercise certain precautions when using the ammeter.

To protect yourself and the ammeter, there is a definite procedure which must be followed when using an ammeter. The first step is to insure that the ammeter you are using is heavy enough for the job. As mentioned above, if the current rating is exceeded, the meter may be damaged.

The second step is to remove power from the circuit to be tested. In battery powered circuits this can be done by removing the battery or by disconnecting one of the battery leads. The purpose of this step is to protect yourself from electrical shock as you connect the ammeter.

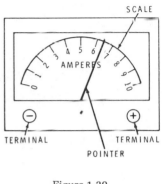

Figure 1-20
Ammeter

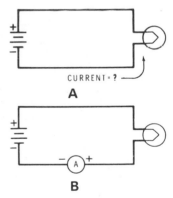

Figure 1-21
Measuring current.

The third step is to break the circuit at the point where the current is to be measured. The circuit must be broken because the ammeter must be placed in series with the circuit.

Fourth, the ammeter is connected to the circuit while *observing polarity*. The ammeter has two terminals labeled negative and positive. Current must flow through the ammeter from the negative terminal to the positive terminal. Thus, the wire from the negative terminal of the battery must lead to the negative terminal of the ammeter. If the ammeter is connected backwards, the pointer will attempt to deflect backwards and may end up bent or broken. *Observing polarity* simply means that the negative terminal of the ammeter is connected to the wire that leads to the negative terminal of the battery. Naturally, the positive terminal of the ammeter is connected to the wire that leads to the positive side of the battery.

Finally, power is reapplied to the circuit and the current is read from the ammeter scale. Figure 1-22 illustrates this step-by-step procedure.

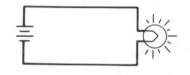

A CIRCUIT IN WHICH CURRENT IS TO BE MEASURED.

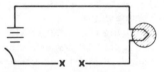

B REMOVE POWER BY DISCON-NECTING ONE SIDE OF BATTERY.

C BREAK CURCUIT AT POINT WHERE CURRENT IS TO BE MEASURED.

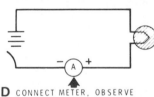

D CONNECT METER, OBSERVE POLARITY.

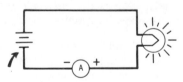

E RESTORE POWER AND READ CURRENT.

Figure 1-22
Procedure for measuring current.

SUMMARY

The following is a summary of the important points discussed in Unit 1. If you have a question on any point presented here, reread that portion of the text which covered this point.

Electronics is that science which controls the behavior of electrons so that some useful function is performed.

Matter is anything which has weight and occupies space.

All matter is composed of one or more of the elements.

A compound is a substance composed of two or more elements.

The smallest particle of a compound is a molecule.

A molecule consists of two or more atoms bound together.

The atom is the smallest particle into which an element can be divided.

There are 92 different types of atoms occurring in nature.

Another dozen or more have been artificially made by man.

Atoms are composed of electrons, protons, and neutrons.

The nucleus contains protons and neutrons.

Electrons orbit the nucleus

The type of atom is determined by the number of electrons, protons, and neutrons.

Electricity is a property that electrons and protons have which causes them to behave in certain predictable ways.

The electron has a negative electrical charge.

The proton has a positive electrical charge.

An electrostatic field surrounds every charged particle.

Coulomb's Law describes the action of charged particles. It states that like charges repel, while unlike charges attract.

An atom has a neutral charge when it contains the same number of electrons and protons.

An atom which has a net electrical charge is called an ion.

Electrical charges can be produced in certain materials by friction.

An electrical charge can be partially transferred from a charged object to an uncharged object by touching the two objects together.

An electrical charge can be induced into a neutral object by bringing a charged object near it.

In electronics, current is defined as the flow of electrical charge from one point to another.

Before an electron can participate in current flow, it must be freed from its atom.

The centrifugal force of the orbiting electron is exactly offset by the attraction of the positive charge in the nucleus.

Electrons are distributed in shells.

The outer shell is called the valence shell.

Valence electrons are the ones important in electronics because they are the ones which can be freed to contribute to current flow.

The number of valence electrons determines if an element is a conductor or an insulator.

A conductor is a substance which has a large number of free electrons.

An insulator is a substance which has very few free electrons.

Most metals are good conductors.

A battery is a two-terminal device which produces an excess of electrons at one terminal and a deficiency of electrons at the other.

Free electrons normally drift around in a random pattern. However, they can be forced to flow in a desired direction.

Current flow is the directed drift of free electrons.

Electrons flow from negative to positive charges.

A schematic diagram uses symbols to represent electronic components.

The unit of electrical charge is the coulomb.

The coulomb is equal to 6.25×10^{18} electrons.

Current is the rate at which electrons flow past a point.

The ampere is the unit of current.

The ampere is equal to one coulomb per second.

A milliampere is one thousandth of an ampere.

A microampere is one millionth of an ampere.

A device for measuring current is the ammeter.

The ammeter must be connected in series with the circuit under test.

Polarity must be observed when connecting an ammeter to a circuit.

Appendix A

SCIENTIFIC NOTATION

In electronics, it is common to deal with both very large and very small numbers. An example of a very large number is the speed at which electricity travels. It travels at the speed of light which is approximately 1,000,000,000 feet per second or about 300,000,000 meters per second. As for very small numbers, consider the size and weight of an electron. It is believed that the electron has a diameter of approximately 0.000 000 000 0022 inch and a weight of about 0.000 000 000 000 000 000 000 000 0009 gram. Sometimes, we perform arithmetic with numbers such as these. To simplify such arithmetic, a shorthand method has been developed to express numbers. This shorthand method is called *scientific notation*. The following programmed instruction sequence will serve as an introduction to scientific notation.

---

1. As mentioned above, scientific notation is a shorthand method of expressing numbers. While any number can be expressed in scientific notation, this technique is particularly helpful in expressing very large and very _____ numbers.

---

2. (small) Scientific notation is based on a concept called powers of ten. Thus, in order to understand scientific notation we should first learn what is meant by powers of _____.

---

3. (ten) In mathematics, a number is raised to a power by multiplying the number times itself one or more times. Thus, we raise 5 to the second power by multiplying 5 times itself. That is, 5 to the second power is $5 \times 5 =$ _____.

---

4. (25) Also, 5 to the third power is the same as saying $5 \times 5 \times 5 =$ _____.

---

5. (125) Thus, 5 can be raised to any power simply by multiplying it times itself the required number of times. For example, $5 \times 5 \times 5 \times 5 = 625$. Consequently, 5 raised to the _____ power is equal to 625.

6. (fourth) The above examples use powers of five. However, any number can be raised to a power by the technique of multiplying it times itself the required number of times. Thus, the powers of two would look like this:

2 to the second power equals $2 \times 2 = 4$

2 to the third power equals $2 \times 2 \times 2 = 8$

2 to the fourth power equals $2 \times 2 \times 2 \times 2 = 16$

2 to the fifth power equals $2 \times 2 \times 2 \times 2 \times 2 = 32$

2 to the sixth power equals $2 \times 2 \times 2 \times 2 \times 2 \times 2 = $ _____

7. (64) In mathematics, the number which is raised to a power is called the *base*. If 5 is raised to the third power, 5 is considered the

_____.

8. (base) The power to which the number is raised is called the *exponent*. If 5 is raised to the third power, then the exponent is 3. In the same way, if 2 is raised to the sixth power, then 2 is the base while 6 is the _____.

9. (exponent) There is a shorthand method for writing "2 raised to the sixth power." It is:

$$2^6$$

Notice that the exponent is written as a small number at the top right of the base. Remember *this* number is the base while *this* number is the exponent.

$$2^6$$

Therefore in the example 3^4, 3 is the _____ while 4 is the

_____.

10. (base, exponent) The number 3^4 is read "3 raised to the fourth power." It is equal to:

$$3 \times 3 \times 3 \times 3 = 81.$$

The number 4^6 is read _____.

11. (4 raised to the sixth power) Scientific notation uses powers of ten. Several powers of ten are listed below:

$$10^2 = 10 \times 10 = 100$$
$$10^3 = 10 \times 10 \times 10 = 1000$$
$$10^4 = 10 \times 10 \times 10 \times 10 = 10,000$$
$$10^5 = 10 \times 10 \times 10 \times 10 \times 10 = 100,000$$
$$10^6 = 10 \times 10 \times 10 \times 10 \times 10 \times 10 = \underline{\qquad}.$$

12. (1,000,000) Multiplication by 10 is extremely easy since all we have to do is add one zero for each multiplication. Another way to look at it is that multiplication by ten is the same as moving the decimal point one place to the right. Thus, we can find the equivalent of 10^2 by multiplying $10 \times 10 = 100$; or, simply by adding a 0 after 10 to form 100; or by moving the decimal point one place to the right to form 10.0, $= 100$. In any event, 10^2 is equal to _____.

13. (100) There is a simple procedure for converting a number expressed as a power of ten to its equivalent number. We simply write down a 1 and after it write the number of zeros indicated by the exponent. For example, 10^6 is equal to 1 with 6 zeros after it. In the same way 10^{11} is equal to 1 with _____ zeros after it.

14. (11) This illustrates one of the advantages of power of ten. It is easier to write and remember 10^{21} than its equivalent number: 1,000,000,000,000,000,000,000. Try it yourself and see if it isn't easier to write 10^{35} than to write its equivalent number of: _____.

31

15. (100,000,000,000,000,000,000,000,000,000,000). In the above examples, we converted a number expressed in powers of ten to its equivalent number. Now let's see how we convert in the opposite direction. Remember the number must be expressed using 10 as the base with the appropriate exponent. The exponent is determined simply by counting the zeros which fall on the right side of the 1. Thus, 1,000,000 becomes 10^6 because there are 6 zeros in the number. In the same way, 10,000,000,000 is expressed as

_____.

16. (10^{10}) To be sure you have the right idea, study each of the groups below. Which group contains an error? _____.

Group A	Group B	Group C
$10^6 = 1,000,000$	$1000 = 10^3$	$10^7 = 10,000,000$
$10^2 = 100$	$10,000 = 10^4$	$10^9 = 1,000,000,000$
$10^9 = 1,000,000,000$	$100 = 10^2$	$10^{11} = 10,000,000,000$

17. (Group C) There are two special cases of powers of ten which require some additional explanation. The first is 10^1. Here the exponent of 10 is 1. If we follow the procedure developed in Frame 13 we find that $10^1 = 10$. That is, we put down a 1 and add the number of zeros indicated by the exponent. Thus $10^1 =$

_____.

18. (10) The other special case is 10^0. Here the exponent is 0. Once again we follow the procedure outlined in Frame 13. Here again we write down a 1 and add the number of zeros indicated by the exponent. However, since the exponent is 0, we add no zeros. Thus, the equivalent number of 10^0 is 1. That is $10^0 =$ _____.

19. (1) Any base number with an exponent of 1 is equal to the base number. Any base number with an exponent of 0 is equal to 1. Thus, $X^1 =$ _____ and $X^0 =$ _____.

20. (X, 1) In the examples given above, the exponents have been positive numbers. For simplicity the plus sign has been omitted. Therefore, 10^2 is the same as 10^{+2}. Also, 10^6 is the same as _____ .

21. (10^{+6}) Positive exponents represent numbers larger than 1. Thus, numbers such as 10^2, 10^7 and 10^{15} are greater than 1 and require _____ exponents.

22. (positive) Numbers smaller than 1 are indicated by negative exponents. Thus, numbers like 0.01, 0.0001, and 0.00001 are expressed as negative powers of ten because these numbers are less than _____ .

23. (1) Some of the negative powers of ten are listed below:
$10^{-1} = 0.1$
$10^{-2} = 0.01$
$10^{-3} = 0.001$
$10^{-4} = 0.0001$
$10^{-5} =$ _____ .

24. (0.00001) A brief study of this list will show that this is simply a continuation of the list shown earlier in frame 11. If the two lists are combined in a descending order, the result will look like this:
$10^6 = 1,000,000.$
$10^5 = 100,000.$
$10^4 = 10,000.$
$10^3 = 1,000.$
$10^2 = 100.$
$10^1 = 10.$
$10^0 = 1.$
$10^{-1} = 0.1$
$10^{-2} = 0.01$
$10^{-3} = 0.001$
$10^{-4} = 0.0001$
$10^{-5} =$ _____

25. (0.00001) We can think of the negative exponent as an indication of how far the decimal point should be moved to the left to obtain the equivalent number. Thus, the procedure for converting a negative power of ten to its equivalent number can be developed. The procedure is to write down the number 1 and move the decimal point to the left the number of places indicated by the negative exponent. For example, 10^{-4} becomes:

0.0001. or 0.0001

Notice that the −4 exponent indicates that the decimal point should be moved _____ places to the _____.

26. (4, left) Up to now we have used powers of ten to express only those numbers which are exact multiples of ten such as 100, 1000, 10,000, etc. Obviously, if these were the only numbers which could be expressed as powers of ten, this method of writing numbers would be of little use. Actually, any _____ can be expressed in powers-of-ten notation.

27. (number) The technique by which this is done can be shown by an example. If 1,000,000 can be represented by 10^6, then 2,000,000 can be represented by 2×10^6. That is, we express the quantity as a number multiplied by the appropriate power of ten. As another example, $2,500,000 = 2.5 \times 10^6$. Also, $3,000,000 =$ _____.

28. (3×10^6) In the same way, we can write 5,000 as 5×10^3. Some other examples are:

$$200 = 2 \times 10^2$$
$$1500 = 15 \times 10^2$$
$$22,000 = 22 \times 10^3$$
$$120,000 = 12 \times 10^4$$
$$1,700,000 = 17 \times 10^5$$
$$9,000,000 = \underline{\hspace{2cm}}$$

29. (9×10^6) By the same token, we can convert in the opposite direction. Thus, 2×10^5 becomes $2 \times 100,000$ or $200,000$. Also, $2.2 \times 10^3 = 2.2 \times 1000 = 2,200$. And, $66 \times 10^4 = $ _____.

30. (660,000) You may have noticed that when we use powers of ten there are several different ways to write a number. For example, 25,000 can be written as 25×10^3 because 25×1000 equals 25,000. However; it can also be written as 2.5×10^4 because $2.5 \times 10,000$ equals 25,000. It can even be written as 250×10^2 since $250 \times 100 = 25,000$. In the same way, 4.7×10^4, 47×10^3, and 470×10^2 are three different ways of writing the number

_____.

31. (47,000) Numbers smaller than one are expressed as negative powers of ten in much the same way. Thus, .0039 can be expressed as 3.9×10^{-3}, 39×10^{-4}, or $.39 \times 10^{-2}$. Also, 6.8×10^{-5}, 68×10^{-6}, and $.68 \times 10^{-4}$ are three different ways of expressing the number

_____.

32. (.000068) As you can see there are several different ways in which a number can be written as a power of ten. Scientific notation is a way of using powers of ten so that all numbers can be expressed in a uniform way. To see exactly what scientific notation is, consider the following examples of numbers written in scientific notation:

$$6.25 \times 10^{18}$$
$$3.7 \ \times 10^6$$
$$4.0 \ \times 10^2$$
$$6.8 \ \times 10^{-4}$$
$$3.9 \ \times 10^{-6}$$
$$2.2 \ \times 10^{-12}$$

Notice that the numbers range from a very large number to an extremely small number. And yet, all these numbers are written in a uniform way. This method of writing numbers is called scientific _____.

33. (notation) The rules for writing a number in scientific notation are quite simple. First, the decimal point is always placed after the first digit on the left which is not a zero. Therefore, the final number will appear in this form: 6.25, 7.3, 9.65, 8.31, 2.0 and so forth. It must never appear in a form such as: .625, 73, 96.5, .831 or 20. Thus, there is always one and only one digit on the _____ side of the decimal point.

34. (left) The second rule involves the sign of the exponent. If the original number is greater than 1, the exponent must be positive. If the number is less than 1, the exponent must be negative. Thus, 67,000 requires a positive exponent but 0.00327 requires a _____ exponent.

35. (negative) Finally, the magnitude of the exponent is determined by the number of places that the decimal point is moved. For example, 39,000.0 is expressed as 3.9×10^4 because the decimal point must be moved 4 places in order to have only one digit to the left of it. Using this rule, 6,700,000,000 is expressed as $6.7 \times$ _____.

36. (10^9) The number 0.00327 is expressed as 3.27×10^{-3}. Here the decimal point is moved 3 places in order to have one digit which is not zero to the left of the decimal. Likewise 0.00027 is expressed as $2.7 \times$ _____.

37. (10^{-4}) To be sure you have the idea look at the groups of numbers below. Which of the following groups contains a number that is not expressed properly in scientific notation? _____

Group A	Group B	Group C
6.25×10^{18}	1.11×10^{11}	6.9×10^{10}
3.75×10^{-9}	-3.1×10^2	3.4×10^7
4.20×10^1	-3.1×10^{-2}	39.5×10^2
7.93×10^0	2.00×10^2	6.0×10^4

38. (Group C) The number 39.5×10^2 is not written in scientific notation because there are two digits on the left side of the decimal point. The minus signs in Group B may have confused you. Although, it has not been mentioned, negative numbers can also be expressed in scientific notation. Thus, a number like $-6,200,000$ becomes -6.2×10^6. All the rules previously stated hold true except that now a _____ sign is placed before the number.

39. (minus) Small negative numbers are handled in the same way. Thus -0.0092 becomes -9.2×10^{-3}. The minus sign before the number indicates that this is a negative number. The minus sign before the exponent indicates that this number is less than

_____.

40. (1) Listed below are numbers which are converted to scientific notation. Which one of these groups contains an error?

_____.

Group A	Group B	Group C
$2,200 = 2.2 \times 10^3$	$119,000 = 1.19 \times 10^5$	$119 = 1.19 \times 10^2$
$32,000 = 3.2 \times 10^4$	$1,633,000 = 1.633 \times 10^6$	$93 = 9.3 \times 10^1$
$963,000 = 9.63 \times 10^5$	$937,000 = 9.37 \times 10^4$	$7.7 = 7.7 \times 10^0$
$660 = 6.6 \times 10^2$	$6,800 = 6.8 \times 10^3$	$131.2 = 1.312 \times 10^2$

41. (Group B) $937,000$ converts to 9.37×10^5 and not to 9.37×10^4. Which of the groups below contains an error?_____.

Group A	Group B	Group C
$0.00037 = 3.7 \times 10^{-4}$	$0.44 = 4.4 \times 10^{-1}$	$.37 = 3.7 \times 10^{-1}$
$0.312 = 3.12 \times 10^{-1}$	$0.0002 = 2.0 \times 10^{-4}$	$.0098 = 9.8 \times 10^{-3}$
$0.068 = 6.8 \times 10^{-2}$	$0.0798 = 7.98 \times 10^{-2}$	$.00001 = 1.0 \times 10^{-5}$
$0.0092 = 9.2 \times 10^3$	$0.644 = 6.44 \times 10^{-1}$	$0.0075 = 7.5 \times 10^{-3}$

42. (Group A) The final number in group A requires a negative exponent. Which of the groups below contains an error?_____ .

Group A	Group B	Group C
$3{,}700{,}000 = 3.7 \times 10^6$	$9440 = 9.44 \times 10^3$	$20 = 2.0 \times 10^1$
$-5{,}500 = -5.5 \times 10^3$	$-110 = -1.1 \times 10^2$	$0.02 = 2.0 \times 10^{-2}$
$0.058 = 5.8 \times 10^{-2}$	$0.0062 = 6.2 \times 10^{-4}$	$-200{,}000 = -2.0 \times 10^5$
$-0.0034 = -3.4 \times 10^{-3}$	$-0.0123 = -1.23 \times 10^{-2}$	$-0.000200 = -2.0 \times 10^{-4}$

43. (Group B) 0.0062 is equal to 6.2×10^{-3}. Match the following:

1. 16   a. 1.6×10^{-3}
2. .0016   b. 1.6×10^4
3. 160,000   c. 1.6×10^0
4. 1.6   d. 1.6×10^1
5. .016   e. 1.6×10^{-2}
6. 16,000   f. 1.6×10^5

44. (1-d, 2-a, 3-f, 4-c, 5-e, 6-b). Another concept that goes hand in hand with powers of ten and scientific notation is metric prefixes. These are prefixes such as *mega* and *kilo* which when placed before a word change the meaning of the word. For example, the prefix *kilo* means *thousand*. When kilo and meter are combined the word kilometer is formed. This word means 1000 meters. In the same way, the word kilogram means _____ grams.

45. (1,000) Since kilo means 1,000 we can think of it as multiplying any quantity times 1000 or 10^3. Thus, kilo means 10^3. Another popular metric prefix is *mega*. *Mega* means *million*. Thus a mega-ton is one million tons or 10^6 tons. In the same way one million volts is referred to as a _____volt.

46. (mega) One thousand watts can be called a kilowatt. Also one million watts can be called a _____.

47. (megawatt) A kilowatt is equal to 10^3 watts while a megawatt is equal to _____ watts.

48. (10^6) Often it is convenient to convert from one prefix to another. For example, since a megaton is 10^6 tons and a kiloton is 10^3 tons, a megaton equals 1000 kilotons. And, since a megaton is one thousand times greater than a kiloton, the kiloton is equal to .001 megaton. Now, consider the quantity 100,000 tons. This is equal to 100 kilotons or _____ megatons.

49. (0.1) Kilo is often abbreviated k. Thus, 100 kilowatts may be expressed as 100 k watt. Mega is abbreviated M. Therefore 10 megawatts may be expressed as _____ watts.

50. (10M) The quantity 5 k volts is 5 kilovolts or 5000 volts. Also, 5 M volts is 5 megavolts or _____ volts.

51. (5,000,000) There are also prefixes which have values less than one. The most used are:

milli	which means *thousandth* (.001) or 10^{-3}, and
micro	which means *millionth* (.000 001) or 10^{-6}.

One thousandth of an ampere is called a milliampere. Also, one thousandth of a volt is called a _____.

52. (millivolt) If a second is divided into one million equal parts each part is called a microsecond. Also, the millionth part of a volt is called a _____.

53. (microvolt) One volt is equal to 1000 millivolts or 1,000,000 microvolts. Or, 1 volt equals 10^3 millivolts and 10^6 microvolts. Expressed another way, 1 millivolt equals .001 volt while 1 microvolt equals .000001 volt. Thus, 1 millivolt equals 10^{-3} volts while 1 microvolt equals _____ volt.

54. (10^{-6}) Powers of ten allow us to express a quantity using whichever metric prefix we prefer. For example, we can express 50 millivolts as 50×10^{-3} volts simply by replacing the prefix milli with its equivalent power of ten. In the same way 50 microvolts is equal to $50 \times$ _____ volts.

55. (10^{-6}) When writing abbreviation for the prefix milli the letter small m is used. A small m is used to distinguish it from mega which used a capital M. Obviously, the abbreviation for micro cannot also be m. To represent micro the Greek letter μ (pronounced mu) is used. Thus, 10 millivolts is abbreviated 10 m volts while 10 microvolts is abbreviated 10 μ volts. Remember, m means 10^{-3} while μ means _____.

56. (10^{-6}) Match the following:

1.	M watt	a.	10^{-3} watts
2.	k watt	b.	10^{-6} watts
3.	m watt	c.	500×10^{-3} watts
4.	μ watt	d.	10^{6} watts
5.	.5 watt	e.	.5k watts
6.	500 watts	f.	10^{3} watts
7.	500,000 watts	g.	.5 M watts
8.	.00005 watts	h.	.05 k watts
9.	50 watts	i.	5 m watts
10.	.005 watts	j.	50 μ watts

57. (1-d, 2-f, 3-a, 4-b, 5-c, 6-e, 7-g, 8-j, 9-h, 10-i) Additional aspects of powers of ten, scientific notation, and metric prefixes will be discussed later.

Unit 2

VOLTAGE

INTRODUCTION

In the previous unit, we saw that a battery produces a force which causes electrons to flow in a closed circuit. We did not learn very much about the force except that it is caused by electrical charges at the battery terminals. That is, the force is produced by an excess of electrons at one terminal and a deficiency of electrons at the other. This force is called an *electromotive force*. Literally, electromotive force means the force which *moves electrons*. The measure of this force is called *voltage*. In fact the terms *electromotive force* and *voltage* are frequently used interchangeably although, strictly speaking, they do have slightly different meanings.

In this unit, we will be discussing voltage or electromotive force in detail. You will learn several ways in which this force is produced and used. You will also learn how voltage is measured and several important laws concerning voltage.

ELECTRICAL FORCE

We have seen that current will not flow in a circuit unless an external force is applied. In the circuits discussed in the previous unit, the force was provided by batteries. The battery changes chemical energy to electrical energy by separating negative charges (electrons) from positive charges (ions). These charges produce the force or pressure which causes electrons to flow and do useful work. This force is given several different names that are used more or less interchangeably. Let's examine the three most popular names and see what each name implies.

Electromotive Force (EMF)

One popular name is *electromotive force* which is abbreviated *emf*. This name is very descriptive since it literally means a force which moves electrons. Thus, emf is the force or pressure which sets electrons in motion. This force is a natural result of Coulomb's law. You will recall that Coulomb's law states that like charges repel while unlike charges attract. The battery, by chemical action, produces a negative charge at one terminal and a positive charge at the other. The negative charge is simply an excess of electrons while the positive charge is an excess of positive ions. If a closed circuit is connected across the battery as shown in Figure 2-1A, a path for electron flow exists between the battery terminals. Free electrons are repelled by the charge on the negative terminal and are attracted by the charge on the positive terminal. The two opposite charges exert a pressure which forces the electrons to flow. Thus, the force or pressure is the result of the attraction of the unlike charges. To summarize, emf is the force which sets electrons in motion in a closed circuit.

Potential Difference

Another name for this force is potential difference. This name is also very descriptive. It describes the characteristics of emf in an open circuit. Emf is the force which causes electrons to move as shown in Figure 2-1A. However, consider the situation shown in Figure 2-1B. Here, electrons cannot flow because the switch is open. Nevertheless, the battery still produces the same pressure or force as before. Thus, the *potential* for producing current flow exists even though no current is presently flowing. As used here, *potential* means the possibility of doing work. If the switch is closed, current flows, the lamp lights, and useful work is done. Therefore, whether a battery is connected into a circuit or not, it has the *potential* for doing work.

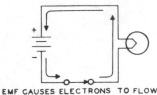

EMF CAUSES ELECTRONS TO FLOW
IN CLOSED CIRCUIT

A

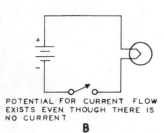
POTENTIAL FOR CURRENT FLOW
EXISTS EVEN THOUGH THERE IS
NO CURRENT

B

Figure 2-1
Emf and potential difference.

Actually, any charge has the potential for doing work. For example, it can move another charge either by attraction or repulsion. Even a single electron can repel another electron. If one electron moves as the result of the action of the other electron, some small amount of work is done. In the battery, we are concerned with two different types of charges rather than a single charge. The electrons at the negative terminal are straining to rush to the positive terminal and cancel out the positive charge there. In the same way, the positive ions at the other terminal are straining to draw the electrons. We call this force *potential difference*. It is the *potential* for doing work that exists between two *different* charges.

The amount of work that can be done must be related in some way to the characteristics of the charges. We can illustrate this point by considering some static charges. Figure 2-2A shows a small negative charge separated from a small positive charge. Let's assume that charge A has an excess of one million electrons while charge B has a deficiency of one million electrons. If a conductor is connected between the two charges, electrons will flow from the negative to the positive charge. The work done here is the moving of electrons. To cancel the two charges, one million electrons will flow from charge A to charge B.

Now consider what happens if the two charges are doubled as shown in Figure 2-2B. Here charge A has an excess of two million electrons while charge B has a deficiency of two million electrons. If a conductor is connected between the two charges, then two million electrons will rush from charge A to charge B. Thus, twice as much work is done.

Figure 2-2
Static charges have
the potential for doing work.

44

However, the magnitude of the charges is not the important consideration. It is the difference between the two charges that is important. Figure 2-3 illustrates that no work can be done if both charges have the same polarity and magnitude. In Figure 2-3A two charges are shown. Each has a negative charge caused by an excess of ten million electrons. How much work is done if these two charges are connected by a conductor? The answer, of course, is that no work is done. Because the two objects have exactly the same charge, no electrons can flow from one to the other Thus, there is no potential for doing work. Figure 2-3B shows that the same is true for equal positive charges.

NO ELECTRONS FLOW BECAUSE
THE TWO CHARGES ARE EQUAL

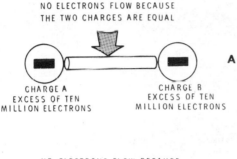

A

CHARGE A
EXCESS OF TEN
MILLION ELECTRONS

CHARGE B
EXCESS OF TEN
MILLION ELECTRONS

NO ELECTRONS FLOW BECAUSE
THE TWO CHARGES ARE EQUAL

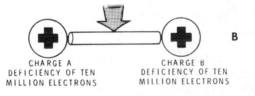

B

CHARGE A
DEFICIENCY OF TEN
MILLION ELECTRONS

CHARGE B
DEFICIENCY OF TEN
MILLION ELECTRONS

Figure 2-3
No potential for doing work
exists between equal charges.

The potential to move electrons exists between any two unlike charges. That is, when two charges are different, electrons will flow from one charge to the other if given the chance. Charges can differ in two ways. First, they can be of opposite *polarity*. This simply means that one is positive and the other is negative as shown in Figure 2-2. Second, they can have different *magnitudes*. For example, Figure 2-4A shows two charges which have the same polarity (negative) but have different magnitudes. Charge A is more negative because it has more excess electrons than charge B. If a conductor is connected between the two charges as shown in Figure 2-4B, electrons will flow from the greater negative charge to the less negative charge. The number of electrons will be exactly the right amount to equalize the two charges. In this example charge A originally has three million excess electrons while charge B has only one million excess electrons. To equalize the two charges, one million electrons will flow from charge A to charge B. Electron flow ceases as soon as the two charges become equal. Notice that the direction of current flow is from the more negative charge to the less negative charge.

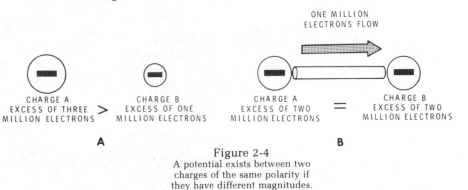

Figure 2-4
A potential exists between two
charges of the same polarity if
they have different magnitudes.

Figure 2-5A illustrates two positive charges of different magnitudes. A potential exists here because electrons will flow if given the chance. Figure 2-5B shows a conductor connecting the two charges. Notice that electrons will flow from the less positive (more negative) to the more positive potential. Again, the number of electrons which flow is the amount necessary to exactly balance the two charges.

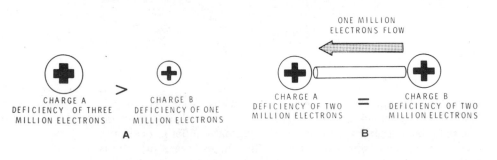

Figure 2-5   Electrons
flow from the less positive
to the more positive charge.

46

Figure 2-6 shows five terminals at various levels of charge. Since no two are at the same charge level, a difference of potential exists between any two terminals. Consequently, if a conductor is placed between any two terminals, electrons will flow until those two charges are balanced. Notice that terminal C has no charge. That is, it contains the same number of electrons as positive ions. Nevertheless, if terminal C is connected to any other terminal, electrons will still flow. If it is connected to one of the negative terminals, electrons will flow into terminal C. If it is connected to one of the positive terminals, electrons will flow from terminal C. Remember that electrons always flow from the more negative to the more positive terminal.

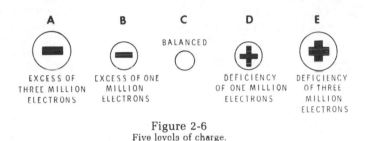

Figure 2-6
Five levels of charge.

Voltage

Another term which is often used interchangeably with emf and potential difference is *voltage*. However, strictly speaking there is a difference between voltage and emf. Voltage is the measure of emf or potential difference. For example, the battery in your car has an emf of 12 volts. The emf supplied by wall outlets is 115 volts while that required by most electric stoves is 220 volts. A large screen color TV receiver produces an emf at one point which may be 25,000 volts or higher. High tension power lines often have a difference of potential as high as 500,000 volts.

The unit of emf or potential difference is the *volt*. At this point it is difficult to visualize exactly how much emf constitutes one volt. However, as you work with electronics, this point will become clearer. One volt is the magnitude of emf which will cause one unit of energy or work to move one coulomb of charge from one point to another. The metric unit of energy or work is the *joule*. This unit is equally difficult to visualize since it is defined in other unfamiliar terms such as newtons. However, joules can be expressed in the more familiar English units. For example, one joule is equal to 0.738 foot-pounds. A foot-pound is the amount of work required to lift one pound one foot. Thus, a joule is approximately the amount of work required to lift 3/4 of a pound one foot off the ground.

47

Using this information, let's return to the volt. One volt is the emf required to cause one joule (or 0.738 foot-pounds) of work to move one coulomb of charge (6.25×10^{18} electrons) from one point to another. To look at it another way, when the movement of one coulomb of charge between two points produces one joule (or 0.738 foot-pounds) of work, the emf between the two points is 1 volt. Later on, after we have discussed resistance, we will define the volt in terms of current and resistance. It will be much easier to visualize then.

The abbreviation of volt is V. Thus, 1.5 volts is abbreviated 1.5V. As with amperes, metric prefixes are attached to indicate smaller and larger units of voltage. Thus, one millivolt equals 1/1000 volt, while one microvolt equals 1/1,000,000 volt. Also, one kilovolt equals 1000 volts while one megavolt equals 1,000,000 volts.

PRODUCING EMF

Emf is produced when an electron is forced from its orbit around the atom. An electric pressure exits between the free electron and the resulting positive ion. Thus, any form of energy which can dislodge electrons from atoms can be used to produce an emf. In no case is energy actually created. It is simply changed to electrical energy from other forms. For example, a battery converts chemical energy to electrical energy while a generator converts mechanical energy to electrical energy.

There are six common methods of producing emf. Each has its own applications. Let's briefly discuss each of these.

Magnetism

This is the most important method of producing electrical power in the world today. At our present level of technology, it is the only method which can produce enough electrical power to run an entire city. Well over 99 percent of all electrical power is produced by this method.

The principle behind this technique is quite simple. When a conductor is moved through a magnetic field, an emf is produced. This is called *magnetoelectricity*. The force of the magnetic field and the movement of the conductor provide the energy necessary to free electrons in the conductor. If the conductor forms a closed loop, then the electrons will flow through the conductor.

The basic requirements are a magnetic field, a conductor, and relative motion between the two. Figure 2-7 shows a simple example. Here the magnetic field is produced by a permanent magnet. The field is represented by the lines drawn from the north to south poles of the magnet. If a conductor is moved up so that it cuts the field as shown in Figure 2-7A, electrons will flow in the direction indicated. The same effect will be obtained if the conductor is held still and the magnet is moved down. All that is required is relative motion between the magnetic field and the conductor.

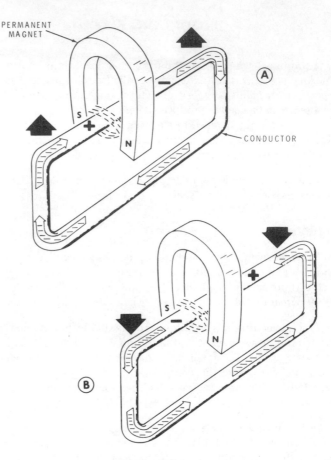

Figure 2-7
Electricity from magnetism.

Figure 2-7B shows that electrons will flow in the opposite direction if the relative motion is reversed. If the conductor is moved up and down in the magnetic field, the electron flow will reverse each time the motion reverses. In generators, a reciprocal motion like this is required. Thus, the current produced alternately flows in one direction then the other. This is known as alternating current or simply AC (or ac). In power generating stations the reciprocal motion occurs 60 times each second. Thus, the power supplied to our homes is often referred to as 60 cycle AC. This is an alternating current which goes through the forward-and-reverse current cycle 60 times each second. In this course, we will not be concerned with alternating current although we will discuss magnetism in some detail later.

Chemical

The next most popular method of generating electricity is by chemical means sometimes called *electrochemistry*. Automobile and flashlight batteries are examples of two typical applications. There are many chemical reactions that will transfer electrons so that an emf is produced.

Figure 2-8 illustrates how a basic battery or cell can be made. A glass beaker is filled with a solution of sulfuric acid and water. This solution is called the *electrolyte*. In the electrolyte, the sulfuric acid breaks down into hydrogen and sulfate. Because of the chemical action involved, the hydrogen atoms give up electrons to the molecules of sulfate. Thus, the hydrogen atoms exist as positive ions while the sulfate molecules act as negative ions. Even so, the solution has no net charge since there are the same number of negative and positive charges.

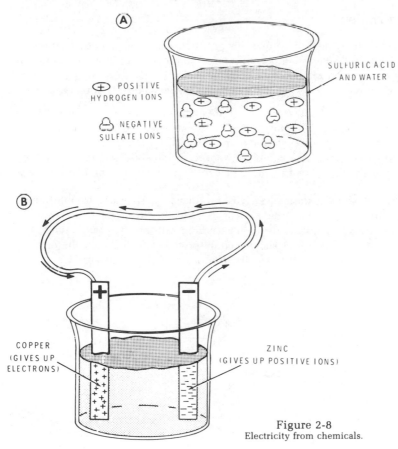

Figure 2-8
Electricity from chemicals.

Next, two bars called *electrodes* are inserted into the solution. One bar is copper while the other is zinc. The positive ions of hydrogen attract the free electrons in the copper. Thus, the copper bar gives up electrons to the electrolyte. This leaves the copper bar with a net positive charge.

The zinc reacts with the sulfate in much the same way. The sulfate molecules have a negative charge. Thus, positive zinc ions are pulled from the bar. This leaves the zinc bar with a surplus of electrons and a net negative charge.

If a conductor is connected between the zinc and copper bars, electrons will flow from the negative to the positive terminal. Because current flow is always in the same direction, we call this direct current or DC (or dc). Compare this to the AC voltage produced by magnetism is Figure 2-7. We will discuss somewhat more practical batteries in detail later in this unit.

Friction

The oldest method known for producing electricity is by friction. We discussed some examples of this in the previous unit. Figure 2-9 shows that a rubber rod becomes negatively charged if rubbed with fur. Also, a glass rod becomes positively charged when rubbed with silk.

You have probably experienced this phenomenum yourself many times. When you scuff your feet across a nylon or wool rug, your shoes develop a charge which is transferred to the body. When you touch a neutral object such as a metal door knob or another person, a discharge occurs. Frequently, there is a tiny arc between your finger and the neutral body.

In many cases, static electricity produced by friction is troublesome or annoying. However, there is a device used in physics laboratories which uses this principle to develop very high voltages. It is called the Van de Graaff generator and some models produce 10 million volts or more. Producing electricity from friction is called the *triboelectric effect*.

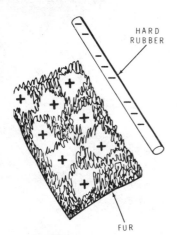

HARD RUBBER

FUR

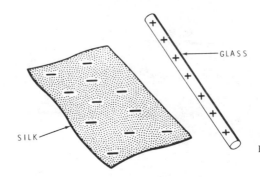

GLASS

SILK

Figure 2-9
Electricity from friction.

Light

Light energy can be converted to electrical energy in large enough quantities to provide limited amounts of power. A familiar example of this is the solar cells frequently used on spacecraft. At present, their cost is too high for commercial use. However, at some time in the future, the price may decline to the point that this type of energy can be used on a much broader scale.

Figure 2-10 shows the construction of one type of solar or photo cell. It consists of some type of photosensitive material sandwiched between two plates which act as electrodes. A photosensitive material is one which develops a charge when it is bombarded by light. Some substances which will do this are cesium, selenium, germanium, cadmium, and sodium. When these materials are struck by light some of the atoms release electrons. This is known as the *photoelectric effect*. In Figure 2-10 light passes through the translucent window and strikes the selenium alloy underneath. Some of the selenium atoms give up electrons and a charge is developed between the two plates. When exposed to sunlight, a single cell can provide a fraction of a volt charge and deliver a few milliamperes of current. When used as a power source, hundreds of the cells are tied together so that they produce usable voltage and current levels.

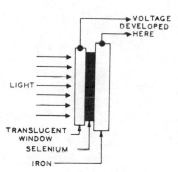

Figure 2-10
Electricity from light.

Pressure

A small electrical charge is developed in some materials when they are subjected to pressure. This is referred to as the *piezoelectric effect*. It is especially noticeable in substances such as quartz, tourmaline, and Rochelle salts all of which have a crystalline structure. Figure 2-11 illustrates how the charge is produced. In the normal structure negative

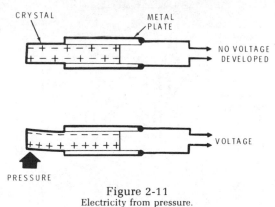

Figure 2-11
Electricity from pressure.

53

and positive charges are distributed so that no overall charge can be measured. However, when the material is subjected to pressure, electrons leave one side of the material and accumulate on the other side. Thus, a charge is developed. When the pressure is relieved, the charges are again distributed so that the net charge disappears.

This effect is put to good use in crystal microphones, phonograph pickups, and precision oscillators. The voltage produced is very small and it must be amplified before it can be used.

Heat

As with most other forms of energy, heat can be converted directly into electricity. The device for doing this is called a *thermocouple*. A thermocouple consists of two dissimilar metals joined together. A typical example is copper and zinc. We have seen that copper will readily give up electrons. This is especially true when the copper is heated. As shown in Figure 2-12, the free electrons from the copper are transferred to the zinc. Thus, the copper develops a positive charge while the zinc develops a negative charge. Since more heat will cause more electrons to transfer, the charge developed is directly proportional to the heat applied. This characteristic allows the thermocouple to be used as a thermometer in areas which are too hot for conventional thermometers. A specific voltage across the thermocouple corresponds to a specific temperature. Therefore, the voltage can be measured and compared to a chart to find the corresponding temperature. The process by which heat is converted directly to electricity is called *thermoelectric effect*.

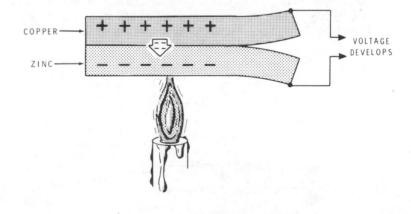

Figure 2-12
Electricity from heat.

54

Effects of Emf

We have seen that an emf can be produced by light, heat, magnetism, pressure, and chemical activity. It is interesting to note that the reverse is also true. That is, an emf can produce light, heat, magnetism, pressure, and chemical activity. The light bulb is an application of light produced by electricity. The toaster and electric stoves are examples where electricity is used to produce heat. When current flows through a wire, the wire is surrounded by a magnetic field. This magnetic field is put to practical use in motors, loud speakers, and solenoids. Recall that a crystal produces a voltage when it is bent or twisted. However, when a voltage is applied to a crystal, the structure bends or twists. Thus, emf can produce pressure. Finally, emf can produce chemical activity. An example of this is the electrolysis of water. When an electric current flows through water, the water is broken down into its component parts of hydrogen and oxygen. Electroplating is another example of chemical activity caused by electricity.

BATTERIES

We discussed one type of battery or cell in the previous section. It consisted of zinc and copper electrodes inserted into an electrolyte of sulfuric acid and water. In this section we will discuss the construction and operation of several more common types of batteries. But first let's discuss the difference between a battery and a cell. A cell is a single unit which contains negative and positive electrodes separated by an electrolyte. A battery is a combination of two or more electrochemical cells. Thus, what we call a flashlight battery is really a cell since it contains only one unit for producing an emf. In spite of this technical definition the word battery is loosely used to describe a single cell.

There are two basic types of cells. One type can be *recharged* and is called a *secondary* cell. The other type can *not* be *recharged* and is called a *primary* cell. All cells and batteries store energy in a chemical form which can be released as electricity.

Dry Cell

Figure 2-13 shows the construction of a flashlight or dry cell. The positive terminal is the steel top at the end of the carbon electrode. The negative terminal is the zinc can or container which holds the rest of the cell. A plastic jacket protects the zinc container and insulates the negative terminal from the positive terminal.

Four popular size batteries.
Courtesy of RCA.

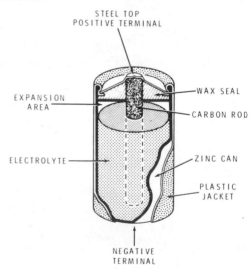

Figure 2-13
Construction of the dry cell.

Even though this type of cell is referred to as a "dry cell," it is not dry on the inside but contains a moist paste. A wax seals off the open end of the zinc container. This prevents any paste from oozing out when the battery is turned upside down or placed on its side. Thus, this type of cell can be used in any position without the electrolyte escaping.

The electrolyte used in this cell is a solution of ammonium chloride and zinc chloride. The electrolyte gradually dissolves the zinc by pulling away positive ions. This process leaves behind an excess of electrons. Thus the remaining zinc acts as the negative electrode. If it were not for the carbon rod, the electrolyte would develop a positive charge by virtue of the positive ions pulled from the zinc. However, the positive charge is neutralized by electrons pulled from the carbon rod. Thus, the carbon rod has a deficiency of electrons which causes a positive charge.

This type of cell is referred to as the Leclanché cell and it produces just over 1.5 volts when new. As it is used, the chemical action slows and the voltage gradually decreases. This type of cell cannot be recharged so it is considered a primary cell. Also, because the paste gradually dries out, the dry cell slowly loses its ability to produce an emf. This occurs even if the battery is not in use. For this reason, the dry cell must be used within about two years of the time it is manufactured. That is, it has a shelf life of about two years.

The voltage delivered by this type of cell is determined strictly by the types of material used as the electrodes and the electrolyte. Thus, the voltage is determined by the chemical reaction and not the size of the cell. For this reason, a small pen light cell produces the same voltage as the much larger D cell. However, the larger battery has a higher current rating. The size D cell can deliver 50 milliamperes of current for approximately 60 hours. A small penlight cell becomes exhausted much sooner at the same current.

Lead-Acid Battery

The prime disadvantage of the dry cell is that it cannot be recharged. The most popular of the cells which can be recharged is the lead-acid cell. Several of these cells are combined to form the lead-acid battery. This is the type of battery found in virtually all automobiles. The principle of the lead-acid cell is illustrated in Figure 2-14.

A positive electrode of lead dioxide (also called lead peroxide) and a negative electrode of spongy lead are immersed in an electrolyte of 8 parts water to 3 parts concentrated sulfuric acid. Sulfuric acid is a combination of sulfate and hydrogen ions. When the cell is discharging, the sulfuric acid combines with both the lead dioxide (positive plate) and the

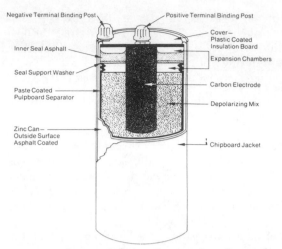

Negative Terminal Binding Post
Positive Terminal Binding Post
Cover—
Plastic Coated
Insulation Board
Inner Seal Asphalt
Expansion Chambers
Seal Support Washer
Carbon Electrode
Paste Coated
Pulpboard Separator
Depolarizing Mix
Zinc Can—
Outside Surface
Asphalt Coated
Chipboard Jacket

Cross section view of a no. 6 dry cell.
Courtesy Union Carbide Corp.

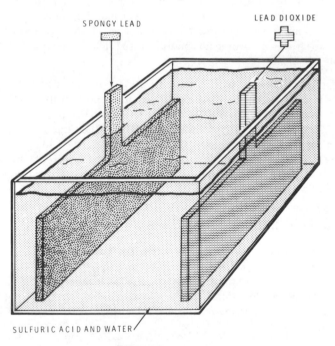

SPONGY LEAD

LEAD DIOXIDE

SULFURIC ACID AND WATER

Figure 2-14
The basic lead-acid cell.

spongy lead (negative plate) converting them to lead-sulfate. The chemical reaction is such that the lead plate develops a negative charge while the lead dioxide plate develops a positive charge. If the discharge continues long enough the lead-sulfate produced covers the two plates to the point that normal operation is impeded. When this happens the cell must be recharged.

Recharging the cell is simply a matter of reversing the current flow through it. This is done by connecting a source of DC voltage greater than that produced by the cell. We have seen that current flow can initiate certain chemical reactions. Here it reverses the chemical action described above. It changes the lead-sulfate in both plates back to sulfuric acid. In doing so, this once again leaves the negative plate pure lead and the positive plate lead dioxide. When the process is complete, the cell is again fully charged.

This type of cell produces an emf of about 2.1 volts. Normally, either 3 or 6 of the cells are combined to form a battery. The cells are connected so that the voltages add. Thus, a three-cell battery has an emf of about 6.3 volts while a six-cell battery has an emf of approximately 12.6 volts. Your automobile has one or the other of these types of batteries.

Because this type of cell can be recharged it is a secondary cell. Also because the electrolyte is a liquid, the lead-acid cell is a wet cell. It must not be laid on its side nor turned upside down. Otherwise, the electrolyte will spill out.

CONNECTING BATTERIES

We have seen that cells can be connected together to increase the voltage or current rating. There are four different ways that cells or batteries can be connected. These are series aiding, series opposing, parallel, and series parallel. Let's examine each of these methods in detail.

Series Aiding Connection

In the 12-volt automobile battery six cells are connected together so that the individual cell voltages add together. In the 6-volt battery, three cells are connected in the same way. This arrangement is called a *series aiding* connection and is shown in Figure 2-15A as it occurs in a three-cell flashlight. The cells are connected so that the positive terminal of the first connects to the negative terminal of the second; the positive terminal of the second connects to the negative terminal of the third; etc. This is a series connection because the same current flows through all three cells. It is an aiding connection because the voltages add together. Since the individual emf of each cell is 1.5 volts, the overall emf is 4.5 volts. The schematic diagram for this connection is shown on the right.

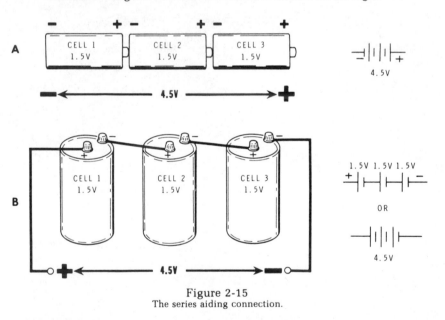

Figure 2-15
The series aiding connection.

Figure 2-15B shows a different type of cell. Here again the three cells are wired together in series. Notice that the voltages add together because between cells the opposite polarity terminals are connected. That is, the negative terminal of the first cell connects to the positive terminal of the next and so on. Thus, the three 1.5-volt cells provide a total emf of 4.5 volts. The schematic symbol is shown on the right.

With the series aiding connection, the total voltage across the battery is equal to the sum of the individual values of each cell. However, the current capacity of the battery does not increase. Since the total circuit current flows through each cell, the current capacity is the same as for one cell.

Series Opposing Connections

The series aiding connection just discussed is extremely important and is widely used. The *series opposing* connection of cells is just the opposite. It has little practical use and is usually avoided. It is mentioned here because an inexperienced person may inadvertently connect cells in this way. The series opposing connection of two cells is shown in Figure 2-16A. Notice that the two cells are connected in series, but like terminals of the cells are connected together. Here the two voltages cancel each other so that the overall emf is 0 volts. Because the two voltages cancel, this arrangement cannot produce current flow.

Figure 2-16B shows another example of a series opposing connection. Here three cells are connected in series but cell number 2 is connected backwards. Consequently, its voltage is subtracted from the voltage of the two cells connected in series aiding. The total voltage for cells 1 and 2 is 0 volts. This leaves the output voltage of cell 3. Therefore, the total output of the three cells is only 1.5 volts.

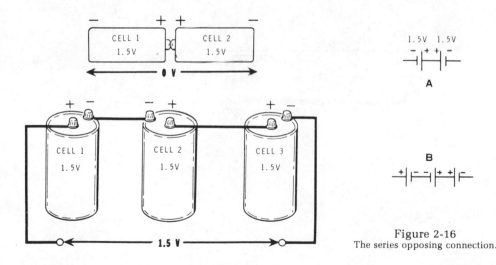

Figure 2-16
The series opposing connection.

Parallel Connection

We have seen that the series aiding connection of cells increases the output voltage but not the current capabilities of the cells. However, there is a way to connect cells so that their current capabilities add together. This is called a parallel connection and is shown in Figure 2-17A. Here, like terminals are connected. That is, all the positive terminals are connected together as are all the negative terminals.

Figure 2-17B shows why the current capacities of the cells are added together. Notice that the total current through the lamp is the sum of the individual cell currents. Each cell provides only one third of the total current. Thus, the total current capacity is three times that of any one cell. However, connecting the cells in this way does not increase the voltage. That is, the total voltage is the same as that for any one cell. If 1.5-volt cells are used, then the total voltage is 1.5 volts.

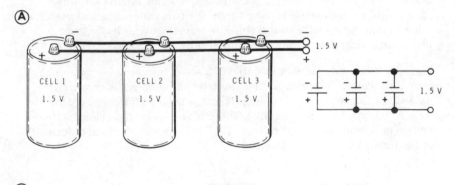

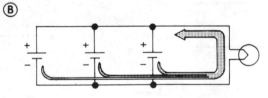

Figure 2-17
The parallel connection.

Series-Parallel Connection

When both a higher voltage and an increased current capacity are required, the cells are connected in series-parallel. For example, suppose we have four 1.5-volt cells and we wish to connect them so that the emf is 3 volts and the current capacity is twice that of any one cell. We can achieve this by connecting the four cells as shown in Figure 2-18. To achieve 3 volts, cells 1 and 2 are connected in series. However, this does not increase the current capacity. To double the current capacity we must connect a second series string (cells 3 and 4) in parallel with the first. The result is the series-parallel arrangement shown.

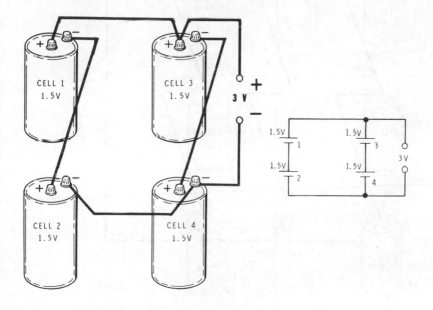

Figure 2-18
The series-parallel connection.

63

To be certain that you have the idea, let's consider another example. Let's suppose that we have a number of identical 1.5 volt cells from which we wish to construct a battery with an emf of 4.5 volts and current capacity three times that of the individual cells. Figure 2-19 shows that nine cells are required. Cells 1, 2, and 3 are connected in a series string to provide 4.5 volts. However, to achieve the higher current, three of these strings must be connected in parallel.

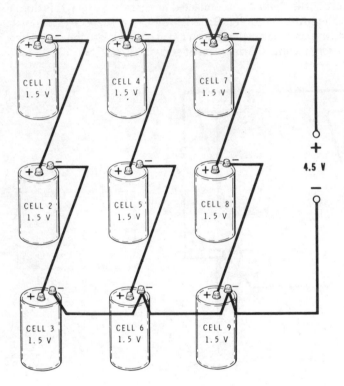

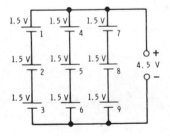

Figure 2-19 Nine cells connected in series-parallel.

VOLTAGE RISES AND VOLTAGE DROPS

In electronics and electrical work there are two kinds of emf or potential difference. Both are expressed in volts but they have somewhat different characteristics. One type of emf is called a voltage rise. The other is called a voltage drop. Let's take a look at the voltage rise first.

Voltage Rise

We have seen that a battery provides an emf or voltage. It does this by chemically producing an excess of electrons at the negative terminal and an excess of positive ions at the positive terminal. When a load is connected across the battery, electrons flow through the load. Each electron which leaves the negative terminal is replaced by an electron from the battery. At the positive terminal each electron arriving from the load cancels one positive ion. However, for each ion that is cancelled, the battery produces a replacement ion. Thus, the voltage between the two terminals remains constant even though electrons are constantly flowing from the negative terminal and into the positive terminal.

Energy is required to move the electrons through the load. The battery gives each electron the energy required to make the trip. We have seen that the energy (in joules) is related to the emf of the battery (in volts) and the number of electrons moved (in coulombs).

The energy comes from the chemical reaction within the battery. This energy has a capacity to do work and the amount of work is determined by the voltage of the battery. After all, it is the emf or voltage of the battery which causes the electrons to flow in the first place. The battery is a source of emf. This type of emf is referred to as a voltage rise. Thus, in an electrical circuit a voltage rise is an emf which is provided by a voltage source.

Earlier, we discussed several different types of voltage sources. The two most common are the generator and the battery. However, solar cells and thermocouples also produce an emf so they are considered voltage sources. Any emf introduced into a circuit by a voltage source is called a *voltage rise*. Thus, a 10-volt battery has a *voltage rise* of ten volts.

Voltage Drop

Electrons which leave the negative terminal of a battery have been given energy by the battery. As the electrons flow through the load, they give up their energy to the load. Most often the energy is given up as heat. However, if the load is a light bulb, both heat and light are given off. The point is that the electrons release to the circuit the energy given to them by the battery.

Since the energy introduced into the circuit is called a voltage rise, the energy removed from the circuit by the load is called a *voltage drop*. A voltage drop is expressed in volts just as in the case with the voltage rise. In fact, the same equation expreses the relationship between volts, joules, and coulombs in both cases. The equation is:

$$\text{volts} = \frac{\text{joules}}{\text{coulombs}}$$

Using this equation, we can determine the voltage drop across a load if we know the energy consumed by the load (in joules) and the number of electrons flowing through the load (in coulombs). For example, let's assume that a light bulb releases 10 joules of energy in one second when a current of two amperes flows through it. With a current of two amperes, two coulombs flow through the bulb each second. Using the above equation, we can determine the voltage drop:

$$\text{volts} = \frac{\text{joules}}{\text{coulombs}}$$

$$\text{volts} = \frac{10 \text{ joules}}{2 \text{ coulombs}}$$

$$\text{volts} = 5$$

Thus, the voltage drop of the light bulb is 5 volts. It is important to realize that this voltage exists between the two terminals of the bulb and it can be measured by a meter. In fact the meter cannot tell the difference between a voltage rise produced by a battery and a voltage drop produced by a load. This is the reason that a battery and a light bulb may both have a rating of 12 volts. In the case of the battery it means that the battery supplies 12 volts. This is a voltage rise. However, for the light bulb, it means that 12 volts is required to make it work. This is a voltage drop.

One difference between the voltage drop and the voltage rise is that the voltage drop occurs only when current flows through the load. Thus, a battery has a voltage rise whether or not it is connected to a circuit. However, a load produces a voltage drop only when current flows through it.

Voltage Drops Equal Voltage Rises

Figure 2-20A shows a 10-volt battery with a light bulb connected across it. The battery provides a voltage rise of 10 volts. As electrons flow through the lamp, a voltage drop is developed across it. Since the lamp consumes the same amount of energy that the battery provides, the voltage drop across the lamp is equal to the voltage rise across the battery. That is, the voltage drop is 10 volts.

In Figure 2-20B, two light bulbs are connected in series across a 10-volt battery. Each bulb drops part of the 10 volts supplied. If the two lamps are identical, then each will drop half of the supplied voltage as shown. If the two lamps are not identical, one bulb will drop more voltage than the other. However, the sum of the voltage drops will always equal the sum of the voltage rises.

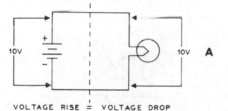

VOLTAGE RISE = VOLTAGE DROP

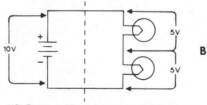

VOLTAGE RISE = VOLTAGE DROPS

Figure 2-20
The voltage drops
are equal to the voltage rises.

To be certain you have the idea, consider the example shown in Figure 2-21A. Here, three batteries are connected series aiding across a single lamp. The sum of the voltage rises is equal to 12 volts. Hence the lamp must drop 12 volts. A final example is shown in Figure 2-21B. Here two 4.5-volt batteries are connected in series with three identical lamps. The total voltage rise in the circuit is 9 volts. Since the lamps are identical each drops one third of the applied voltage or 3 volts. Notice that once again, the sum of the voltage rises equals the sum of the voltage drops.

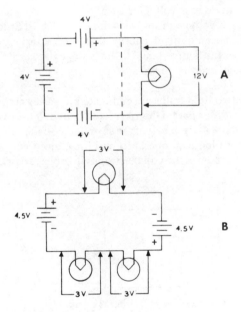

Figure 2-21   The sum of the voltage drops equals the sum of the voltage rises.

CONCEPT OF GROUND

One of the most important points in the study of electricity and electronics is the concept of ground. Originally ground was just what the name implies — the earth. In fact, in some countries the name earth is used instead of ground. Earth is considered to have zero potential. Thus, ground or earth is the reference point to which voltages are most often compared. Many electrical appliances in your home are grounded. This is especially true of air conditioning units, electric clothes dryers, and washing machines. Often this is done by connecting a heavy wire directly to a cold water pipe which is buried deep in the earth (ground). In other cases, a third prong on the power plug connects the metal frame to ground. The purpose of this is to protect the user in case a short circuit develops in the appliance. It also places the metal parts of different appliances at the same potential so that you are not shocked by a difference in potential between two appliances. This type of ground is sometimes called earth ground.

However, there is a slightly different type of ground used in electronics. For example, a certain point in a small transistor radio is called ground although the radio does not connect to earth in any way. This is the concept of ground with which we will be primarily concerned in this course. In this case, ground is simply a zero reference point within an electric circuit. In most larger pieces of electronic equipment the zero reference point or ground point is the metal frame or chassis on which the various circuits are constructed. All voltages are measured with respect to this chassis.

In your automobile, the chassis or metal body of the automobile is considered ground. If you look closely at the straps leaving the battery you will see that one wire connects directly to the metal frame of the car. This point is considered to be ground as is every other point on the metal frame.

In electronics, ground is important because it allows us to have both negative and positive voltages. Up to now we have been concerned only with relative voltages between two points. For example, a 6-volt battery has an emf between its two terminals of 6 volts. We do not think of this as +6 volts or −6 volts but rather simply 6 volts.

However, the concept of ground allows us to express negative and positive voltages. Remember ground is merely a reference point which is considered zero or neutral. If we assume that the positive terminal of a 6-volt battery is ground, then the negative terminal is 6 volts more negative. Thus, the voltage at this terminal with respect to ground is −6 volts.

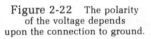

A

B

C

D

Figure 2-22 The polarity
of the voltage depends
upon the connection to ground.

On the other hand, if we assume that the negative terminal of the battery is ground, then the positive terminal with respect to ground is +6 volts. Notice that the battery can produce −6 volts or +6 volts depending on which terminal we assign to ground.

Many small electronic devices such as calculators, transistor radios, etc. do not have metal frames. Instead, all components are mounted on a printed circuit board. Here ground is nothing more than an area of copper on the board. However, as before, all voltages are measured with respect to this point. In this case, ground is simply a common reference which is a handy starting point for measuring voltages.

The schematic symbol for ground is shown in Figure 2-22A. Figure 2-22B shows how it is used in the circuit. Point A is at ground or zero potential. Now, since this is a 10-volt battery, point B is at a plus 10 volt potential with respect to ground. We say that point B is ten volts above ground or that the voltage at this point with respect to ground is +10 volts.

Figure 2-22C shows why ground is so important. Here the same battery is shown but with the positive terminal connected to ground. That is, the positive terminal is the zero volt point in this circuit. Because the negative terminal is ten volts more negative, the voltage at point A with respect to ground is −10 volts. Thus, we can use the battery as a +10-volt source or as a −10-volt source depending on where we connect ground.

Another example is shown in Figure 2-22D. Here two batteries are connected in series, with the ground connection between them. Thus, the zero reference is at point B. Since the top battery has an emf of 10 volts, the voltage at point C with respect to ground is +10 volts. The lower battery has an emf of 6 volts. Because the positive terminal is connected to ground, the emf at point A with respect to ground is −6 volts.

We sometimes loosely speak of the voltage at a particular point. Actually, voltage is always the measure of the potential difference between *two* points. Thus, in Figure 2-22D, when we speak of the voltage at point A, what we really mean is the voltage between point A and ground.

70

MEASURING VOLTAGE

The device used for measuring voltage is called the voltmeter. There are many different types of voltmeters in use today. Most use a mechanical meter movement like the one shown in Figure 2-23A. Here the voltage is read on a scale behind the moving pointer. Another type, called the digital voltmeter, is becoming increasingly popular. It is shown in Figure 2-23B. Here the voltage is displayed as numerals. This type of meter is more accurate and is easier to read. Generally, it is also more expensive.

IM 2260

The low-priced Heathkit IM-2260 Portable Digital Multimeter features attractive styling, a wide measurement range and highly accurate readings.

IM 5284

Figure 2-23
Two types of voltmeters.

71

Regardless of the type of voltmeter, certain precautions must be taken to ensure accurate readings. To begin with, voltage is always measured *between two points*. The schematic symbol for the voltmeter is shown in Figure 2-24A. Notice that one of the leads is marked negative while the other is marked positive. As with the ammeter discussed earlier, polarity must be observed when using the voltmeter. This means that the negative lead must go to the more negative of the two points across which the voltage is to be measured.

Fortunately, the voltmeter is much easier to use than the ammeter. With the voltmeter, the circuit under test need not be broken nor disturbed in any way. To measure the voltage between two points, we merely touch the two leads of the voltmeter to the two points. However, we do have to observe polarity. Figure 2-24B shows how the voltmeter is connected to measure the voltage drop across the lower lamp. Notice that the negative lead is connected to the more negative point. Notice also that the meter is connected directly across the lower bulb.

In this case, the meter is measuring a voltage drop. If the switch is opened so that current flow stops, the voltage drop disappears and the meter reading falls to zero.

Figure 2-24C shows a different circuit with the meter connected to measure the voltage rise of the lower battery. Once again polarity is observed and the meter is connected directly across the component. Here the meter is measuring a voltage rise, rather than a voltage drop. Therefore, the voltage remains constant when the switch is opened. Notice that the voltage rise does not depend on current flow.

A couple of precautions should be taken when using a voltmeter. First, we should always make certain that the voltage we are going to measure is not higher than the meter can measure. If it is, the high voltage may damage the meter. Also, we should be certain the meter is on the proper range. For example, we should not attempt to measure 100 volts with the meter set to the 1 volt range. This, too, may damage the meter. When we are unsure of the value of voltage being measured, we should make our first measurement with the meter on a high range. This will prevent the unknown voltage from *pegging* the meter.

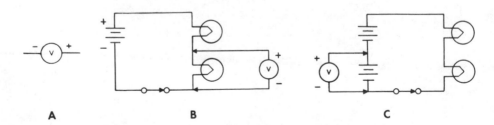

A   **B**   **C**

Figure 2-24
Connecting the voltmeter.

For our own safety there are some other precautions which we should observe. For example, we should hold the meter leads only by the insulated portions. Otherwise, we may receive an electrical shock. As we have seen, in most electronic devices, voltages are measured with respect to ground. Therefore, when working on electronic equipment, it is a good idea to connect one lead to ground and leave it there. This way only one hand is required to make voltage measurements. The other hand can be held away from the equipment. This greatly reduces our chances of receiving an electrical shock since there is no complete path for current flow through our body.

SUMMARY

The following is a point-by-point summary of Unit 2. If you are in doubt about any of the facts given here, you should review the appropriate portion of the text.

Emf is the force which moves electrons. This force is a natural result of Coulomb's law.

Potential difference is another name for the same force. It represents the potential for moving electrons.

Emf or potential difference exists between any two charges which are not exactly alike.

A difference of potential exists between any uncharged body and any charged body.

An emf exists between two unequal positive charges, between unequal negative charges, and between any negative charge and any positive charge.

Voltage is the measure of emf or potential difference.

The unit of voltage is the volt.

One volt is equal to one joule of energy or work per coulomb of transferred charge.

The joule is the metric unit of work and is equal to 0.738 foot pounds.

The millivolt is 1/1000 volt. The microvolt is 1/1,000,000 volt. The kilovolt is 1000 volts. The megavolt is 1,000,000 volts.

Emf can be produced in several different ways. The most common method uses magnetism and mechanical motion. Other methods use chemical reactions, friction, light, pressure, or heat.

A cell is an electrochemical device which contains negative and positive electrodes separated by an electrolyte.

A battery is a combination of two or more cells.

There are two basic types of cells — the primary cell and the secondary cell.

The primary cell cannot be recharged.

The secondary cell can be recharged.

The dry cell uses a paste-like electrolyte and can be used in any position.

The wet cell uses a liquid electrolyte and can be operated only in the upright position.

The most common type of dry cell uses an electrolyte of ammonium chloride and zinc chloride. The positive electrode is carbon and the negative electrode is zinc. It produces an emf of about 1.5V when new. It is a primary cell.

The most common type of wet cell is used in the lead-acid battery.

The electrolyte is sulfuric acid and water. The positive electrode is lead dioxide and the negative electrode is pure lead. It produces an emf of about 2.1V. It is a secondary cell which can be recharged repeatedly.

The voltage produced by a cell is determined by the chemical reaction and not by the size of the cell. However, current capacity is determined by the size.

When two or more cells are connected series aiding, the individual voltages add. Thus, a 6 volt battery can be formed by connecting four 1.5 volt cells in the series aiding arrangement.

When two cells are connected series opposing, the voltage produced by one cell is subtracted from that produced by the other. This connection is normally avoided.

When two or more cells are connected in parallel, the output voltage is the same as that for any one cell. That is, the voltages do not add. However, the current capacity does increase.

Cells can be connected in a series-parallel arrangement so that both the voltage and the current capacity increase.

A voltage can exist at two different points in a circuit: Where emf is produced and where energy is used.

A voltage which exists because emf is being produced is called a voltage rise.

A voltage which exists because energy is being used is called a voltage drop.

A voltage rise exists with or without current flow.

A voltage drop exists across a component only when current flows through the component.

If one joule of energy is consumed by a load when one coulomb of charge flows through it, then the voltage drop across the load is one volt.

In a circuit in which current is flowing, the sum of the voltage drops is equal to the sum of the voltage rises.

Ground is the name given to the point in a circuit that is used as the zero reference. Often this is the metal chassis or frame of the electronic device.

Voltages may be negative or positive with respect to ground. A battery can be connected as a negative voltage rise or as a positive voltage rise depending on how it is connected to ground.

The device used to measure voltage is called the voltmeter.

Voltage is measured between two points that have a difference of potential.

We must observe polarity when connecting a voltmeter to a circuit.

The voltmeter is connected across the voltage rise or voltage drop to be measured. Thus, we need not break the circuit under test when measuring voltage.

Unit 3

RESISTANCE

INTRODUCTION

In this unit, you will study resistance. Resistance is that property which opposes current flow. All materials have this property to some extent. Some materials such as glass and rubber offer a great deal of opposition to current flow. They allow almost no current to flow through them. Thus, they are said to have a very high resistance. Other materials such as silver and copper offer very little opposition to current flow. Therefore, they have a very low resistance.

This unit deals with the characteristics of resistance. It describes what resistance is, how it is measured, and how it is used.

BASIC CONCEPTS

If a short length of *copper* is connected across a battery, a great deal of current will flow — so much, in fact, that the battery will quickly discharge. If the same length of *rubber* is connected across the same battery, no measurable current will flow. Obviously, then copper and rubber have very different characteristics when it comes to passing current flow.

The difference stems from the atomic structure of the two materials. In copper, there are a great number of free electrons drifting aimlessly through the spaces between the atoms. When an emf is applied, a great number of electrons are free to move. Thus, a high current flow results. In rubber, the situation is different. Here very few electrons drift around free between the molecules. Therefore, an emf can cause very few electrons to move. The result is little or no current flow.

Copper offers little opposition to current flow while rubber offers a great deal of opposition. This opposition to current flow is called resistance. Copper has very low resistance while rubber has very high resistance. Consequently, the resistance of a material is determined largely by the number of free electrons in the material.

The Ohm

The unit of resistance is the *ohm*. This unit is named for Georg Simon Ohm a German physicist who discovered the relationship between voltage, current, and resistance.

An ohm may be defined in several different ways. Originally, the ohm was defined as the resistance of a column of mercury which is 106.3 centimeters long and 1 square millimeter in cross sectional area. Once the ohm was defined in this way, it allowed anyone with the proper equipment to construct a 1 ohm standard. Unfortunately, this amount of resistance is hard to visualize.

It may be helpful to think in terms of something with which we are more familiar. For example, a length of number 22 copper wire that is 60 feet long has a resistance of one ohm.

The most common way to define the ohm is in terms of voltage and current. One ohm is the amount of resistance which will allow one ampere of current to flow in a circuit to which one volt of emf is applied. Or stated another way, if one volt causes one ampere of current in a circuit then the resistance of the circuit is one ohm. Figure 3-1 illustrates these three ways of defining the ohm.

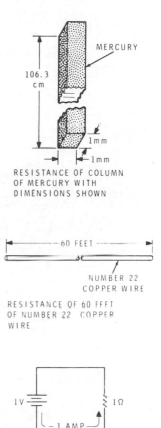

RESISTANCE OF COLUMN OF MERCURY WITH DIMENSIONS SHOWN

RESISTANCE OF 60 FEET OF NUMBER 22 COPPER WIRE

RESISTANCE WHICH ALLOWS ONE AMPERE TO FLOW WHEN EMF IS 1 VOLT

Figure 3-1   The ohm

79

The Greek letter omega (Ω) is commonly used to represent ohms. Thus, 1 ohm may be written 1 Ω. Also, one thousand ohms may be written as 1000 Ω, 1 kilohms, or as 1 KΩ. Finally, one million ohms may be written as 1,000,000 Ω, 1 megohms, or as 1 MΩ. In electronics, the letter R is used to represent resistance. Thus, in the shorthand of electronics the statement "the resistance is ten ohms" may be written as an equation: "R = 10 Ω."

Resistivity

We cannot directly compare the resistances of two substances because the resistance depends on the shape and size of the substance as well as the temperature. However, every substance has a property called *specific resistance* or *resistivity* which can be compared directly. The resistivity of a substance is defined as the resistance of a one-foot length of wire of the substance. Furthermore, the wire must be exactly 0.001 inch (1 mil) in diameter and the temperature must be exactly 20° centigrade. These requirements standardize the shape, size, and temperature of the substance so that only the atomic structure determines the resistance.

The resistivity of several substances is shown in Figure 3-2. Notice that silver has the lowest resistivity while copper is a close second. Near the bottom of the list are glass and rubber. Silver and copper are the best conductors while glass and rubber are two of the best insulators.

SUBSTANCES	RESISTIVITY IN OHM PER MIL. FOOT AT 20°C
SILVER	9.9
COPPER	10.4
GOLD	15.3
ALUMINUM	17.0
NICHROME	660.0
GLASS	10^{16}
RUBBER	10^{20}

Figure 3-2
Resistivity of common materials

As we have seen, both conductors and insulators are important in electronics. Conductors have many free electrons so they conduct current very easily. Thus, they are used to carry electricity from one place to another. Most metals are good conductors. Thus, metals such as silver, copper, gold, aluminum, tungsten, zinc, brass, platinum, iron, nickel, tin, steel, and lead are all good conductors.

Insulators or nonconductors are substances which have few free electrons. They have very high values of resistivity. These substances are used to prevent electrical connection. Most wires are coated with an insulator so that they do not accidentally short out when used to carry electricity. Some examples of insulators are glass, rubber, plastic, mica, and dry air.

To summarize, it is the resistivity of a material which determines if the material is a conductor or an insulator. Resistivity is the resistance of a specific size and shape of the material at a specific temperature.

Conductance

Sometimes it is more convenient to think in terms of how well a material conducts currents rather than to think in terms of how well it opposes current. Because of this, a property called conductance is often used. Conductance is just the opposite of resistance. It is defined as the ease with which a substance passes current flow. Mathematically, conductance is the reciprocal of resistance. This simply means that conductance is equal to the number 1 divided by the resistance. Or stated as an equation:

$$\text{Conductance} = \frac{1}{\text{Resistance}}$$

The letter G is used to represent conductance. Therefore, the equation can be written:

$$G = \frac{1}{R}$$

The unit of conductance is the mho, pronounced "moe." Notice that this is ohm spelled backwards. The mho is the reciprocal of the ohm. Therefore,

$$\text{mhos} = \frac{1}{\text{ohms}}$$

A resistance of 1 ohm equals a conductance of 1 mho. However, a resistance of 2 ohms equals a conductance of 1/2 or 0.5 mhos. Also, if the resistance is 1000 ohms or 1 kilohm, the conductance is 0.001 mhos or 1 millimho.

In most cases it is more convenient to think in terms of ohms (resistance) rather than in terms of mhos (conductance). Therefore, we will be primarily concerned with resistance in this course. However, it is important to remember the mho because a key characteristic of the vacuum tube and the field-effect transistor is given in mhos.

Factors Determining Resistance

The single most important factor in determining resistance is the resistivity of the material. However, three other factors are also important. These are the length, the cross-sectional area and the temperature of the material. This is the reason that these three variables had to be carefully defined in order to determine resistivity. Let's discuss each of these in more detail.

Length. A 60-foot length of number 22 copper wire has a resistance of about 1 ohm. A 120-foot length of the same wire has a resistance of about 2 ohms. Thus, if we double the length of the wire, we also double the resistance. In other words, the resistance of a conductor is directly proportional to its length. In fact, with any material,the greater the length, the higher the resistance will be. The reason for this is that the electrons must travel further through the resistance medium. Thus, if the length doubles, the resistance doubles; if the length triples, the resistance triples; and so forth.

Cross-Sectional Area. The cross-sectional area of a conductor is determined by its thickness or its diameter. We have seen that good conductors have a large number of free electrons. In fact the more free electrons per unit of length, the better the conductor will be. Obviously then, a large diameter conductor has more free electrons per unit of length than a small diameter conductor of the same material. Therefore, large diameter conductors have less resistance than small diameter conductors.

All other things being equal, the resistance of a substance is inversely proportional to its cross-sectional area. If the cross-sectional area doubles, the resistance drops to one half its former value. Also, if the area triples, the resistance drops to one third.

Temperature. When defining resistivity, the length and cross-sectional area are carefully defined because they affect resistance. The temperature is also carefully defined for the same reason. That is, with most materials, the resistance changes if the temperature changes. With changes in length and cross-sectional area, we know exactly how the resistance will change. Not only that, but all materials change resistance in the same way. However, with changes in temperature, this is not the case. Not all materials change resistance in the same direction or by the same amount when temperature changes.

In most materials, an increase in temperature causes an increase in resistance. Materials which respond in this way are said to have a *positive temperature coefficient*. If a material has a positive temperature coefficient, its resistance increases as temperature increases and decreases as temperature decreases.

A few substances such as carbon have a *negative temperature coefficient*. This means that their resistance decreases as temperature increases. There are also materials whose resistances do not change at all with temperature. These materials are said to have a *zero* or *constant* temperature coefficient.

In most simple circuits, the temperature coefficients of the components are not critical and are simply ignored. However, in some circuits the temperature coefficients are important and they must be considered in the design.

However, the temperature characteristics are not always troublesome. A device called a *thermistor* uses these characteristics to great advantage. A thermistor is a special type of resistor which can achieve a large change in resistance for a small change in temperature. Such devices normally have a negative temperature coefficient. In many thermistors the resistance value can drop to one half its former value for a temperature rise of 20 degrees centigrade. Thermistors are often used in temperature sensing circuits and as protective devices in other types of circuits.

RESISTORS

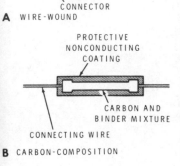

CERAMIC FORM
RESISTANCE WIRE

METAL
CONNECTOR
A WIRE-WOUND

PROTECTIVE
NONCONDUCTING
COATING

CARBON AND
BINDER MIXTURE

CONNECTING WIRE

B CARBON-COMPOSITION

C SCHEMATIC SYMBOL

Figure 3-3 Types of
resistors and their schematic symbol.

A resistor is an electronic component which has a certain specified resistance. Of course, other types of components also have some resistance. But the resistor is designed specifically to introduce a desired amount of resistance into a circuit.

Wire-Wound Resistors

We have seen that copper has a resistivity of about 10 ohms per mil-foot. We could wrap a one foot length of 1 mil diameter copper wire on an insulated form and attach contact leads as shown in Figure 3-3A. This would form a 10 ohm wire-wound resistor. The process for producing practical wire-wound resistors is a little more involved but the idea is the same.

The resistance wire used is generally a nickel-chromium alloy called nichrome which has a much higher resistivity than copper. The form is often a ceramic tube. After the leads are attached the entire resistor is covered with a hard protective coating. This type of resistor is often used in high current circuits where relatively high amounts of power must be dissipated. The resistance range can vary from less than an ohm to several thousand ohms. The wire-wound technique is also used to produce precise value resistors. Such precision values are required in meter circuits.

Wire-wound resistors are formed by wrapping resistance wire around a ceramic core. The unit is sealed by a hard plastic insulating material.
(Courtesy IRC Division of TRW Inc.)

84

Carbon-Composition Resistors

The element carbon is neither a good conductor nor a good insulator. Instead, it falls in-between and is called a *semiconductor*. This makes carbon ideal as a material for resistors. By combining carbon granules and a powdered insulating material in various proportions a wide range of resistor values are possible.

Figure 3-3B shows the construction of this type of resistor. Granules of carbon and a binder material are mixed together and shaped into a rod. Wire leads are inserted and the package is sealed with a non-conducting coating. The schematic symbol for the resistor is shown in Figure 3-3C.

Carbon-composition resistors are inexpensive and are the most common type used in electronics. Generally, they are used in low current circuits where they do not have to dissipate large amounts of power. Values can vary from 10 ohms or less to 20 megohms or more. The resistance value is indicated by color bands around the resistor.

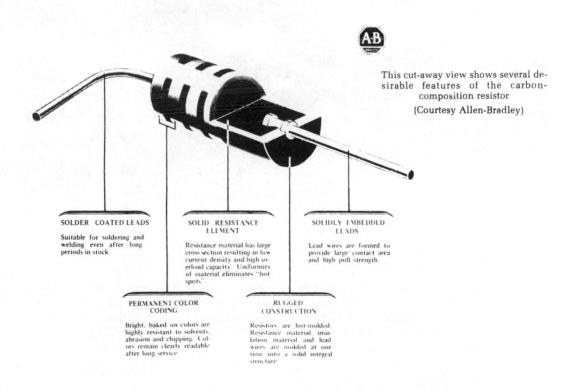

This cut-away view shows several desirable features of the carbon-composition resistor

(Courtesy Allen-Bradley)

SOLDER COATED LEADS

Suitable for soldering and welding even after long periods in stock.

SOLID RESISTANCE ELEMENT

Resistance material has large cross section resulting in low current density and high overload capacity. Uniformity of material eliminates "hot spots".

SOLIDLY EMBEDDED LEADS

Lead wires are formed to provide large contact area and high pull strength.

PERMANENT COLOR CODING

Bright, baked on colors are highly resistant to solvents, abrasion and chipping. Colors remain clearly readable after long service.

RUGGED CONSTRUCTION

Resistors are hot-molded. Resistance material, insulation material and lead wires are molded at one time into a solid integral structure.

Deposited-Film Resistors

Another type of resistor which is becoming increasingly popular is the film-type resistor. The construction of this type of resistor is shown in Figure 3-4. In these devices, a resistance film is deposited on a nonconductive rod. Then the value of resistance is set by cutting a spiral groove through the film. This changes the appearance of the film to that of a long, flat ribbon spiralled around the rod. The groove adjusts the length and width of the ribbon so that the desired value is achieved. Several different types of deposited film resistors are available. The most common type is the carbon-film resistor. Here a carbon film is deposited on a ceramic rod. Several metal-film types are also available. One uses a nickel-chromium (nichrome) film on an aluminum oxide rod. Another uses a tin-oxide film on a glass rod.

Figure 3-4
Deposited film resistor
(Courtesy IRC Division of TRW Inc.)

Resistor Ratings

Resistors have three very important ratings: resistance (in ohms), tolerance (in percent) and wattage (in watts). If we know what to look for, these ratings can usually be determined simply by examining the resistor. Let's look at each of these ratings in more detail.

Resistance. We have discussed resistance in some detail already. We have seen that it is determined by the length, the cross-sectional area, and the resistivity of the material used. With wire-wound resistors, the value is normally written somewhere on the resistor. However, with carbon-composition and film resistors the value is usually indicated by color bands.

Tolerance. The resistance is rarely the exact value indicated on the resistor. It would be extremely difficult and expensive to make the resistor the exact value indicated. For this reason, resistors have a tolerance rating. For example, a 1,000 ohm resistor may have a tolerance of ± 10 percent. Ten percent of 1000 is 100. Therefore, the actual value of the resistor can be anywhere from 900 ohms (1000 − 100) to 1100 ohms (1000 + 100).

Tolerances of ± 5 percent, ± 10 percent, and ± 20 percent are common for carbon composition resistors. Precision resistors often have a tolerance of ± 1 percent or less. Generally, the lower the tolerance, the more the resistor costs.

Wattage. Wattage rating refers to the maximum amount of power or heat that the resistor can dissipate without burning up or changing value. As shown in Figure 3-5, the larger the physical size of the resistor, the more power it can dissipate and the higher the wattage. Carbon composition resistors generally have fairly low wattage ratings. Ratings of 2 watts, 1 watt, 1/2 watt, and 1/4 watt are the most common. Wire wound resistors can have much higher wattage ratings. A rating as high as 250 watts is not too uncommon for wire wound resistors.

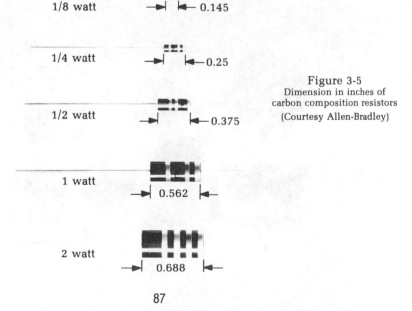

1/8 watt — 0.145

1/4 watt — 0.25

1/2 watt — 0.375

1 watt — 0.562

2 watt — 0.688

Figure 3-5
Dimension in inches of
carbon composition resistors
(Courtesy Allen-Bradley)

87

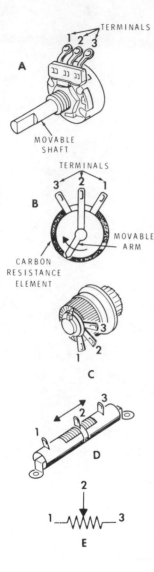

Variable Resistors

The volume controls on our TV receivers and radios are examples of variable resistors. These are resistors whose values can be changed simply by rotating a shaft.

Figure 3-6A shows the construction of a variable carbon resistor. Figure 3-6B is a rear-view of the inside of the device which shows how the resistance is changed. A flat, circular strip of carbon is mounted between the two end terminals. A contact which moves along the resistance element connects to the center terminal. This arm is attached to the moveable shaft. If the arm is moved in the direction shown by the arrow, the resistance between terminals 1 and 2 increases. Notice that, simultaneously, the resistance between terminals 2 and 3 decreases. This type of variable resistor is called a *potentiometer* or simply a *pot*. A potentiometer has *three* terminals.

A slight variation of this arrangement is called a rheostat. A rheostat has only two terminals. Thus, this arrangement can be changed to a rheostat by removing or not using terminal 3. Notice that the resistance between terminals 1 and 2 can still be varied without terminal 3. However, without the third terminal the flexibility of the device is greatly reduced.

Wire-wound potentiometers are also common. Many have the same outward appearance as the potentiometer shown in Figure 3-6A. However, the internal construction is slightly different. As shown in Figure 3-6C, resistance wire is wound around an insulating core. A contact arm moves along the bare wire changing the resistance between the center and outside terminals.

Figure 3-6
Variable resistors.

Potentiometers are available in a wide variety of shapes and sizes

(Courtesy Allen-Bradley)

Another type of variable resistor is shown in Figure 3-6D. This type is sometimes called a sliding contact resistor. It is used in high-power applications where the resistance value must be initially set or occasionally reset. The resistance value is changed by moving the sliding contact along the bare resistance wire. Figure 3-6E shows the schematic diagram of the variable resistor.

CONNECTING RESISTORS

Resistors are often connected in series, in parallel, and in series-parallel combinations. In order to analyze and understand electronic circuits, we must be able to compute the total resistance of resistor networks.

Resistors In Series

As mentioned earlier a series circuit is one in which the components are connected end to end as shown in Figure 3-7A. Notice that the same current flows through all components. The current in the circuit must flow through all three resistors one after the other. Therefore, the total opposition to current flow is the total of the three resistances.

An example is shown in Figure 3-7B. Here three resistors are connected in series. The total resistance is called R_T and is the resistance across the entire series circuit. R_T is found by adding the individual resistor values together. That is:

$$R_T = R_1 + R_2 + R_3$$
$$R_T = 10 \ \Omega + 20 \ \Omega + 30 \ \Omega$$
$$R_T = 60 \ \Omega$$

The three resistors in series have the same opposition to current flow as a single 60 ohm resistor.

This example used three resistors. However, the same principle holds true for any number of series resistors. Thus, in Figure 3-7C a 1 K ohm resistor in series with a 5 K ohm resistor has a total resistance of 6 K ohms or 6000 ohms. The circuit shown in Figure 3-7D uses six resistors. Some of the values are given in kilohms while others are given in ohms. To avoid confusion, we can convert all values to ohms. R_1, R_3, and R_5 are already expressed in ohms. R_2 is equal to 1 K ohms or 1000 ohms. R_4 is 1.2 K ohms or 1,200 ohms. And, R_6 is 3.3 K ohms or 3300 ohms. Thus, the total resistance R_T is found by adding:

$$R_T = R_1 + R_2 + R_3 + R_4 + R_5 + R_6$$
$$R_T = 500 \ \Omega + 1000 \ \Omega + 750 \ \Omega + 1200 \ \Omega + 600 \ \Omega + 3300 \ \Omega$$
$$R_T = 7350 \ \Omega \text{ or } 7.35 \ K\Omega$$

A

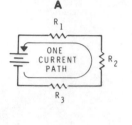

B

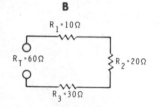

C

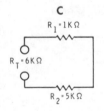

D

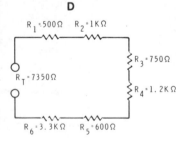

Figure 3-7
Resistors in series

Resistors In Parallel

In parallel circuits components are connected across each other so that there are two or more paths for current flow. Figure 3-8A is an example.

To see how resistors in parallel act, let's add switch S1 in series with R_2 so that we can switch R_2 in and out of the circuit. The resulting circuit is shown in Figure 3-8B. With S_1 open, a certain current will flow through R_1. The amount of current is determined by the resistance of R_1 and the applied voltage. Because there is only one path for current flow, the current through R_1 is the total circuit current.

Now let's close the switch and see what happens. The current through R_1 remains unchanged since neither the resistance of R_1 nor the applied voltage has changed. However, an additional current now flows through R_2. Thus, the total current provided by the battery has increased. If R_2 has the same resistance as R_1, both resistors will offer the same amount of opposition to current flow. Thus, the current through R_1 will equal the current through R_2. In this case, the current provided by the battery doubles when R_2 is switched in parallel with R_1.

Regardless of the size of R_1 and R_2, the total current provided by the battery will always increase when R_2 is placed in parallel with R_1 because a second current path is created. Obviously then, the total opposition to current flow decreases since more current now flows. Thus, when one resistor is placed in parallel with another, the total resistance decreases. Let's assume that R_1 and R_2 are equal. When the switch is closed, the total current doubles. Thus, the total resistance has dropped to one half its former value.

There is a simple formula for finding the total resistance of two resistors in parallel. The formula is:

$$R_T = \frac{R_1 \times R_2}{R_1 + R_2}$$

Figure 3-8C shows an example. Here a 15 ohm resistor (R_1) is in parallel with a 10 ohm resistor (R_2). Let's find the total resistance R_T.

$$R_T = \frac{R_1 \times R_2}{R_1 + R_2}$$

$$R_T = \frac{15\ \Omega \times 10\ \Omega}{15\ \Omega + 10\ \Omega}$$

$$R_T = \frac{150}{25}$$

$$R_T = 6\ \Omega$$

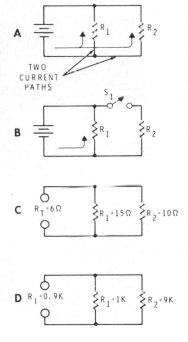

Figure 3-8
Resistors in parallel

91

The total resistance of the circuit is the same as that of a 6 ohm resistor.

Figure 3-8D shows another example. Here a 1 K ohm resistor is in parallel with a 9 K ohm resistor.

$$R_T = \frac{R_1 \times R_2}{R_1 + R_2}$$

$$R_T = \frac{1000 \times 9000}{1000 + 9000}$$

$$R_T = \frac{9,000,000}{10,000}$$

$$R_T = 900 \text{ ohms or } 0.9 \text{ K ohms}$$

The above formula is normally used when two resistors are in parallel. However, it can also be used with three or more resistors. For example, Figure 3-9A shows four resistors in parallel. Using the above formula an equivalent resistance for R_1 and R_2 can be found. Then using the formula again an equivalent resistance for R_3 and R_4 can be found. Finally, the formula can be applied to the two equivalent resistances so that the total resistance can be found. However, this method involves using the formula three different times.

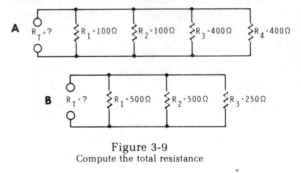

Figure 3-9
Compute the total resistance

There is a better formula to use when more than two resistors are connected in parallel. The formula is:

$$R_T = \frac{1}{\dfrac{1}{R_1} + \dfrac{1}{R_2} + \dfrac{1}{R_3} + \dfrac{1}{R_4} + \cdots \dfrac{1}{R_n}}$$

Using this formula for the circuit shown in Figure 3-9A, we can find the total resistance (R_T):

$$R_T = \cfrac{1}{\cfrac{1}{100} + \cfrac{1}{100} + \cfrac{1}{400} + \cfrac{1}{400}}$$

$$R_T = \cfrac{1}{.01 + .01 + .0025 + .0025}$$

$$R_T = \cfrac{1}{.025}$$

$$R_T = 40 \text{ ohms}$$

Figure 3-9B shows an example using only three resistors. Let's find R_T.

$$R_T = \cfrac{1}{\cfrac{1}{R_1} + \cfrac{1}{R_2} + \cfrac{1}{R_3}}$$

$$R_T = \cfrac{1}{\cfrac{1}{500} + \cfrac{1}{500} + \cfrac{1}{250}}$$

$$R_T = \cfrac{1}{.002 + .002 + .004}$$

$$R_T = \cfrac{1}{.008}$$

$$R_T = 125 \text{ ohms}$$

Equal Resistors In Parallel

Sometimes two or more resistors which have the same value are placed in parallel. There is a simple rule which covers this situation. When all the resistors in parallel have the same value, the total resistance can be found by dividing that value by the number of resistors in parallel. For example, Figure 3-10A shows two 20 ohm resistors connected in parallel. Our rule states that:

$$R_T = \frac{\text{value of one resistor}}{\text{number of resistors in parallel}}$$

$$R_T = \frac{20\ \Omega}{2}$$

$$R_T = 10\ \Omega$$

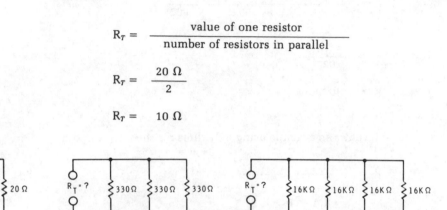

Figure 3-10 When all resistors have same value, R_T equals value divided by number in parallel

Likewise, when three resistors of the same size are connected in parallel, the total resistance is one third the value of one resistor. Thus, in Figure 3-10B:

$$R_T = \frac{330\ \Omega}{3}$$

$$R_T = 110\ \Omega$$

In Figure 3-10C:

$$R_T = \frac{16K\ \Omega}{4}$$

$$R_T = 4\ K\ \Omega$$

94

Series-Parallel Connections

In many circuits a parallel circuit is connected in series with one or more resistors as shown in Figure 3-11A. Even so, the total resistance is easy to calculate using the formulas shown earlier. The procedure is to compute an equivalent resistance for the parallel circuit. Then this equivalent resistance is added to the series resistance values.

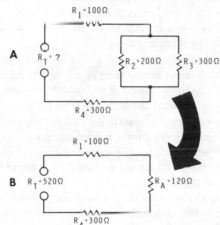

Figure 3-11
Simplifying the series-parallel circuit.

For example, we would first find the equivalent resistance for the parallel network made up of R_2 and R_3. Let's call this equivalent resistance R_A. Using the formula for a two resistor parallel network, we find:

$$R_A = \frac{R_2 \times R_3}{R_2 + R_3}$$

$$R_A = \frac{200 \times 300}{200 + 300}$$

$$R_A = \frac{60,000}{500}$$

$$R_A = 120 \text{ ohms}$$

Now we can substitute R_A for the parallel network as shown in Figure 3-11B. Once we have the circuit in this simplified form, we use the formula for finding the total resistance in a series circuit.

$$R_T = R_1 + R_A + R_4$$
$$R_T = 100 \ \Omega \ + \ 120 \ \Omega \ + \ 300 \ \Omega$$
$$R_T = 520 \text{ ohms}$$

95

SUMMARY

The following is a summary of Unit 3.

Resistance is the opposition to current flow. The unit of resistance is the ohm. The ohm may be expressed in terms of voltage and current. It is equal to the amount of resistance which will allow one ampere of current to flow when an emf of 1 volt is applied. The Greek letter omega is used to represent the ohm.

Resistivity is defined as the resistance of a one mil-foot length of a material at 20 degrees centigrade. A mil-foot is a one foot length of wire which is one thousandth of an inch in diameter. The resistivity determines if a material is an insulator or a conductor.

Most metals have a very low resistivity and are therefore good conductors. Some examples are silver, copper, aluminum, nickel, iron, lead, and gold. Other materials have values of resistivity billions of times higher. These materials are good insulators. Examples are glass, rubber, mica, and plastics.

The resistance of a material is determined not only by its resistivity but also by its size and shape. The resistance is directly proportional to the length of a conductor but inversely proportional to the cross-sectional area.

Temperature also affects resistance to some extent. If the resistance increases with temperature, the substance is said to have a positive temperature coefficient. If the resistance decreases with temperature, the substance is said to have a negative temperature coefficient.

There are three popular types of fixed resistors. These are: carbon-composition, wire-wound, and deposited-film. Carbon-composition resistors are the most common and least expensive while wire-wound resistors are the most expensive. Wire-wound resistors generally have relatively low resistance values but they can have high power ratings. Deposited-film resistors fill the gap between composition and wire-wound resistors. They can be made more precise than composition resistors and they are cheaper than wire-wound resistors.

The resistance, tolerance, and wattage of most resistors can be determined by close examination of the resistor. With many wire-wound resistors this information is written on the body of the resistor. With the composition and film resistors, the resistance is indicated by three color bands. The tolerance is indicated by a fourth band and the wattage is indicated by the physical size of the resistor.

Not all resistors have a fixed value; some are variable. A resistor whose value can be changed by rotating a shaft is called a potentiometer if it has three terminals, but is called a rheostat if it has only two terminals.

Resistors may be connected in series or in parallel. When connected in series their resistance values add. Thus, if two 1000 ohm resistors are connected in series, the total resistance is 2000 ohms.

When resistors are connected in parallel, the total resistance decreases. If two 1000 ohm resistors are connected in parallel, the total resistance becomes 500 ohms.

The device used for measuring resistance is the ohmmeter. Most ohmmeters have a nonlinear scale with 0 at one end and infinity at the other. A range switch is provided so that a wide range of resistance values can be accurately measured. A zero adjust is also included on most meters. This control must be readjusted each time the range is changed.

All electronic components have some resistance value. The light bulb works only because of the resistance of its filament. It is this resistance that produces the heat which causes the filament to glow. The fuse is another example. Here a fragile length of resistance wire is designed to burn in two when the current rating is exceeded.

The thermistor is a special type of resistor whose resistance value changes with temperature.

97

Unit 4

OHM'S LAW

INTRODUCTION

Ohm's law is the most important and most basic law of electricity and electronics. It defines the relationship between the three fundamental electrical quantities: current, voltage, and resistance. Fortunately, the relationship between these three quantities is quite simple. Several implications of this relationship have already been discussed in the previous units. Therefore, some of the information presented in this unit will not be entirely new to you.

OHM'S LAW

Ohm's law defines the way in which current, voltage, and resistance are related. We examined this relationship to some extent in the previous units. Now, let's examine it more closely and in a systematic way.

Determining Current

Ohm's law states that *current is directly proportional to voltage and inversely proportional to resistance.* Figure 4-1A will help illustrate this point. The source of voltage is the battery. Voltage is the force which causes current to flow. Therefore, the higher the voltage, the higher the current will be. Conversely, the lower the voltage, the lower the current will be. This assumes that the resistance remains constant. However, as we have seen, the current is also determined by the resistance. Resistance is the opposition to current flow. Assuming that the voltage is constant, the higher the resistance, the lower the current will be. Also, the lower the resistance, the higher the current will be.

These facts can be summarized by a single formula:

$$\text{current} = \frac{\text{voltage}}{\text{resistance}}$$

Or, stated in terms of the units of current, voltage, and resistance:

$$\text{amperes} = \frac{\text{volts}}{\text{ohms}}$$

Figure 4-1
Current is determined
by voltage and resistance.

100

When used in formulas, single letters of the alphabet are used to represent current, voltage, and resistance. Resistance is represented by the letter R. Voltage may be represented either by the letter V (for voltage) or the letter E (for EMF). In this course we will use the letter E to represent EMF or voltage. Current is represented by the letter I. While this may seem a little illogical, this convention is used throughout electronics. If we substitute the letters I, E, and R for the quantities current, voltage, and resistance respectively; the formula for current becomes:

$$I = \frac{E}{R}$$

This formula may be used to find current in any circuit in which the voltage and resistance are known.

Figure 4-1B shows a circuit in which the values of voltage and resistance are given. To determine the current we merely substitute the known values into the formula:

$$I = \frac{E}{R}$$

$$I = \frac{10 \text{ volts}}{5 \text{ ohms}}$$

$$I = 2 \text{ amperes}$$

Notice that we simply divide 5 into 10 and receive a final answer of 2. Anytime we divide ohms into volts the answer is expressed in amperes. Thus, the current in the circuit shown in Figure 4-1B is 2 amperes.

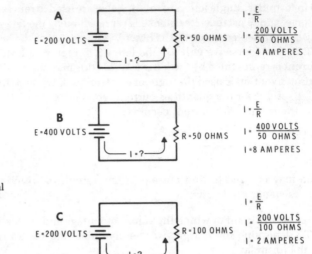

A

E = 200 VOLTS, R = 50 OHMS, I = ?

$I = \dfrac{E}{R}$

$I = \dfrac{200 \text{ VOLTS}}{50 \text{ OHMS}}$

I = 4 AMPERES

B

E = 400 VOLTS, R = 50 OHMS, I = ?

$I = \dfrac{E}{R}$

$I = \dfrac{400 \text{ VOLTS}}{50 \text{ OHMS}}$

I = 8 AMPERES

Figure 4-2
Current is directly proportional
to voltage and inversely
proportional to resistance

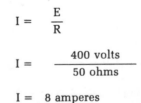

C

E = 200 VOLTS, R = 100 OHMS, I = ?

$I = \dfrac{E}{R}$

$I = \dfrac{200 \text{ VOLTS}}{100 \text{ OHMS}}$

I = 2 AMPERES

Let's look at another example. Figure 4-2A shows another circuit in which the value of E and R are given. Solving for I, we find that the current is:

$$I = \frac{E}{R}$$

$$I = \frac{200 \text{ volts}}{50 \text{ ohms}}$$

$$I = 4 \text{ amperes}$$

Let's see what happens to the current, if we double the applied voltage as shown in Figure 4-2B.

$$I = \frac{E}{R}$$

$$I = \frac{400 \text{ volts}}{50 \text{ ohms}}$$

$$I = 8 \text{ amperes}$$

Notice that when the voltage is doubled, the current also doubles. We expect this because current is *directly proportional* to voltage.

Now, let's return to Figure 4-2A. What happens if we double the resistance and hold the voltage constant? This situation is shown in Figure 4-2C. The current becomes:

$$I = \frac{E}{R}$$

$$I = \frac{200 \text{ volts}}{100 \text{ ohms}}$$

$$I = 2 \text{ amperes}$$

Thus, when we double the resistance, the current is reduced to one-half its former value. We expect this since current is *inversely proportional* to resistance.

Let's consider another example. How much current flows through a 3 K ohm resistor when it is connected across a 9 volt battery? The easiest way to solve this problem is to convert 3 K ohms to 3000 ohms. Thus:

$$I = \frac{E}{R}$$

$$I = \frac{9 \text{ volts}}{3000 \text{ ohms}}$$

$$I = 0.003 \text{ amperes}$$

You will recall that 0.003 amperes is another way of saying 3 milliamperes. Because resistance values are often given in kilohms, it is convenient to be able to work problems without converting kilohms to ohms. In current problems, when volts are divided by kilohms, the result is expressed in milliamperes.

For example, how much current flows when a lamp with a resistance of 2.4 kilohms is connected across a 120 volt line? The current in milliamperes is:

$$I = \frac{E}{R}$$

$$I = \frac{120 \text{ volts}}{2.4 \text{ K ohms}}$$

$$I = 50 \text{ milliamperes}$$

103

We can check this by converting 2.4 kilohms to 2400 ohms and solving as we did earlier:

$$I = \frac{.E}{R}$$

$$I = \frac{120 \text{ volts}}{2400 \text{ ohms}}$$

$$I = 0.05 \text{ amperes}$$

You will recall that 0.05 amperes is the same as 50 milliamperes.

Resistance is often given in megohms. For example, what is the current when a 5 megohm resistor is connected across a 25 volt battery? We can convert 5 megohms to 5,000,000 ohms and solve for current in amperes:

$$I = \frac{E}{R}$$

$$I = \frac{25 \text{ volts}}{5,000,000 \text{ ohms}}$$

$$I = 0.000\ 005 \text{ amperes}$$

This is 5 microamperes. Thus, when we divide volts by megohms the answer is in microamperes. That is:

$$I = \frac{E}{R}$$

$$I = \frac{25 \text{ volts}}{5 \text{ megohms}}$$

$$I = 5\ \mu \text{ amps.}$$

To summarize, the basic formula for current is:

$$I = \frac{E}{R}$$

If R is given in ohms, we can think of the equation as:

$$\text{amperes} = \frac{\text{volts}}{\text{ohms}}$$

104

However, if R is given in kilohms, we can think of the equation as:

$$\text{milliamperes} = \frac{\text{volts}}{\text{kilohms}}$$

Finally, if R is given in megohms, we can think of the equation as:

$$\text{microamperes} = \frac{\text{volts}}{\text{megohms}}$$

Finding Voltage

We have seen that the formula for current is:

$$I = \frac{E}{R}$$

By *transposing* this equation, we can develop a formula for voltage. Transposing simply means changing the equation from one form to another. Since we are interested in finding voltage, we must change the formula so that E is on one side of the equation by itself. This is easy to do if we remember a basic algebra rule for transposing equations. The rule states that we can multiply or divide both sides of the equation by any quantity without changing the equality. The current equation is:

$$I = \frac{E}{R}$$

Multiplying both sides by R we have:

$$I \times R = \frac{E \times R}{R}$$

Notice that R appears in both the numerator and the denominator of the fraction on the right side of the equation. You will recall from basic mathematics that the two R's in the fraction can be cancelled like this:

$$I \times R = \frac{E \times \cancel{R}}{\cancel{R}}$$

This leaves:

$$I \times R = E$$

We can reverse the two sides of the equation without changing the equality. Reversing the two sides, we find the basic equation for voltage:

$$E = I \times R$$

In other words, voltage is equal to current times resistance. Generally, the times sign ($\times$) is omitted so that the formula is written simply:

$$E = IR$$

A

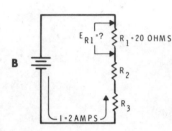

Figure 4-3A shows a circuit in which the resistance and current are known. Let's find the voltage:

$$E = IR$$

$$E = 0.5 \text{ amperes} \times 125 \text{ ohms}$$

$$E = 62.5 \text{ volts}$$

Notice that amperes times ohms gives volts.

B

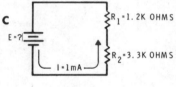

Figure 4-3B shows a slightly different problem. Here, we wish to find the voltage drop across R_1. We will call this voltage E_{R1}. We know that $R_1 = 20$ ohms and that the current through R_1 is 2 amperes. Consequently, we can find the voltage drop across R_1:

$$E_{R1} = I \times R_1$$

$$E_{R1} = 2 \text{ amperes} \times 20 \text{ ohms}$$

$$E_{R1} = 40 \text{ volts}$$

C

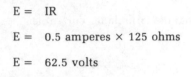

Figure 4-3C shows two known resistance values connected in series. The current is given but the battery voltage is not. To find the battery voltage we must multiply the total resistance (R_T) times the current. The total resistance is calculated by adding the two series resistances. Thus,

$$R_T = R_1 + R_2$$

$$R_T = 1.2 \text{ K ohms} + 3.3 \text{ K ohms}$$

$$R_T = 4.5 \text{ K ohms}$$

D

Figure 4-3
Finding voltage

Once the total resistance is known, the voltage can be computed. However, notice that the resistance is given in kilohms while the current is given in milliamperes. We could change kilohms to ohms and milliamperes to amperes then solve for volts as we did above. However, this is unnecessary since milliamperes multiplied by kilohms equal volts. Thus,

$$E = I \times R_T$$

$$E = 1 \text{ milliampere} \times 4.5 \text{ kilohms}$$

$$E = 4.5 \text{ volts}$$

Figure 4-3D shows a partial schematic in which the resistance is given in megohms, the current is given in microamperes, and the voltage drop is unknown. Once again, it is unnecessary to convert to ohms and amperes because megohms multiplied by microamperes gives volts. Thus, to find the voltage drop across R, we simply multiply:

$$E = I \times R$$

$$E = 2 \mu A \times 6.8 \text{ M}\Omega$$

$$E = 13.6 \text{ volts}$$

Finding Resistance

We can transpose the current formula or the voltage formula to produce a resistance formula. For example, the voltage formula is:

$$E = IR$$

Dividing both sides by I we get,

$$\frac{E}{I} = \frac{IR}{I}$$

The I's in the fraction on the right cancel:

$$\frac{E}{I} = \frac{\not{I}R}{\not{I}}$$

So the formula becomes:

$$\frac{E}{I} = R$$

Reversing the equation we have the formula for resistance:

$$R = \frac{E}{I}$$

This states that resistance is equal to voltage divided by current. Or,

$$ohms = \frac{volts}{amperes}$$

Using this formula, we can find the resistance in any circuit in which the voltage and current are known. Figure 4-4A shows such a circuit. Solving for R, we find that:

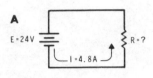

A

$$R = \frac{E}{I}$$

$$R = \frac{24V}{4.8A}$$

$$R = 5 \text{ ohms}$$

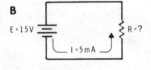

B

Figure 4-4B shows another example. Here the current is given in milliamperes while the emf is expressed in volts. When milliamperes are divided into volts the result is in kilohms. Thus:

$$R = \frac{E}{I}$$

$$R = \frac{15 \text{ V}}{5 \text{ mA}}$$

$$R = 3 \text{ kilohms}$$

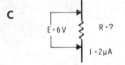

C

Figure 4-4
Finding resistance.

This can be easily proven by converting 5 mA to 0.005 A and solving for R:

$$R = \frac{E}{I}$$

$$R = \frac{15 \text{ V}}{0.005 \text{ A}}$$

$$R = 3000 \text{ ohms}$$

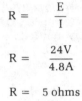

Finally, in Figure 4-4C we wish to find the resistance which drops 6 volts when the current is 2 microamps. When microamps are divided into volts the result is megohms. Thus,

$$R = \frac{E}{I}$$

$$R = \frac{6 \text{ V}}{2 \, \mu \text{ A}}$$

$$R = 3 \text{ megohms}$$

Once again this can be proven by converting 2 μ A to 0.000 002 amps and dividing:

$$R = \frac{E}{I}$$

$$R = \frac{6 \text{V}}{0.000 \ 002 \ \text{A}}$$

$$R = 3,000,000 \text{ ohms}$$

Summary

Ohm's law may be expressed by three different formulas:

$$I = \frac{E}{R}$$

$$E = I R$$

$$R = \frac{E}{I}$$

In Figure 4-5A the three quantities are diagrammed in a way that may help you to remember these three equations. To use the diagram, cover the quantity for which you wish to find the equation. For example, in Figure 4-5B we wish to find the current so the quantity I is covered. The quantity covered is the left hand side of the equation. The remaining two symbols represent the right hand side of the equation. Notice that the remaining quantities are

$$\frac{E}{R}$$

Therefore,

$$I = \frac{E}{R}$$

Figures 4-5C and D show how to find the formulas for resistance and voltage.

109

A

B

$$I - \frac{E}{R}$$

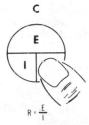

C

$$R - \frac{E}{I}$$

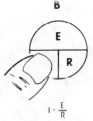

D

$$E - I R$$

Figure 4-5
The three forms of Ohm's law.

Some equally handy diagrams are shown in Figure 4-6. These indicate how the quantities are grouped with and without metric prefixes. In Figure 4-6A no metric prefixes are used with any of the quantities. In any Ohm's law problem, if two of the quantities are given in the units shown in Figure 4-6A then the third quantity is given in the third unit. For example, if resistance is given in ohms while voltage is given in volts, then current will be in amperes.

In Figure 4-6B, emf is still given in volts. However, current is given in milliamperes while resistance is given in kilohms. It is important to remember that these three quantities go together. Thus, if resistance is given in kilohms while emf is given in volts, the current will be in milliamperes. Or, if current is in milliamperes while resistance is in kilohms, the emf will be in volts.

Figure 4-6C shows that a similar relationship exists between volts, microamps, and megohms. For example, if emf is given in volts and current is given in microamperes, resistance will be in megohms.

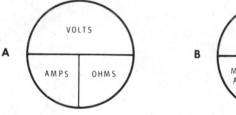
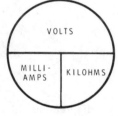
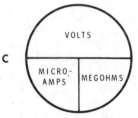

Figure 4-6
How the metric prefixes are related.

POWER

In addition to the three basic electrical quantities (current, voltage, and resistance) a fourth quantity is also very important. This quantity is called *power*. Power is defined as the rate at which work is done. In other words, power refers to the amount of work done in a specific length of time.

Work and Power

In an earlier unit you learned that a *joule* is equal to the amount of work done by one volt of emf in moving one coulomb of charge. Notice that time does not enter into this definition. Thus the same amount of work is done whether the charge is moved in one second or in one hour.

Unlike *work*, power is concerned with time.. Therefore, power is the measure of joules per unit of time. The unit of power is the *watt*. It is named in honor of James Watt who pioneered the development of the steam engine. The watt is equal to *one joule per second.*

The English unit of work (the foot-pound) is easier to visualize than the joule. If one pound is lifted vertically by one foot then one foot-pound of work is done. The joule is equal to 0.738 foot-pounds. Therefore, the watt is equal to 0.738 foot-pounds per second. The unit of mechanical power in the English system is the horsepower. The horsepower is equal to 550 foot-pounds per second. In other words, if 550 pounds are raised one foot in one second then one horsepower is expended. In terms of watts, the horsepower is equal to 746 watts. Or, stated another way, the watt is equal to 1/746 or 0.00134 horsepower.

Power, Current, and Voltage

We have seen that the watt is equal to one joule per second. That is, the watt is the work done in one second by one volt of emf in moving one coulomb of charge. If one coulomb of charge flows in one second then the current is one ampere. Thus, one watt is the amount of power used in a circuit when one ampere of current flows as the result of one volt of applied emf.

Power is directly proportional to both current and voltage. Thus, the formula for power is:

$$\text{power} = \text{current} \times \text{voltage}$$

Or, stated in terms of the units of the three quantities:

$$\text{watts} = \text{amperes} \times \text{volts}$$

111

When used in equations, the symbol P is used to represent power. Thus, the formula for power is:

$$P = IE$$

As with Ohm's law, there are two other useful forms of this equation. The first expresses voltage in terms of current and power. It is found by rearranging the equation:

$$P = IE$$

Dividing both sides by I:

$$\frac{P}{I} = \frac{IE}{I}$$

Cancelling the two I's on the right side:

$$\frac{P}{I} = \frac{\cancel{I}E}{\cancel{I}}$$

Reversing the two sides of the equation:

$$E = \frac{P}{I}$$

This states that voltage (in volts) is equal to power (in watts) divided by current (in amperes).

Another useful form of the equation is:

$$I = \frac{P}{E}$$

See if you can derive this formula from the original power formula given above.

To summarize, power, voltage and current are related by the following formulas:

$$P = IE, \quad E = \frac{P}{I}, \quad \text{and } I = \frac{P}{E}$$

Now, let's work some sample problems using these formulas.

What is the power dissipated in a circuit in which an emf of 50 volts is applied and a current of 3.2 amperes flows?

$$P = I E$$

$$P = 3.2 \text{ amperes} \times 50 \text{ volts}$$

$$P = 160 \text{ watts}$$

How much current flows through a 75 watt light bulb which is connected across a 120 volt power line?

$$I = \frac{P}{E}$$

$$I = \frac{75 \text{ watts}}{120 \text{ volts}}$$

$$I = 0.625 \text{ amperes}$$

What is the voltage drop across a light bulb that dissipates 60 watts when the current through the bulb is 0.5 amperes?

$$E = \frac{P}{I}$$

$$E = \frac{60 \text{ watts}}{0.5 \text{ amperes}}$$

$$E = 120 \text{ volts}$$

Power Dissipation in Resistors

In resistors and most other electronic components power is dissipated in the form of heat. In some cases, the heat produced is a desired result. For example, the purpose of the resistance element in a toaster, heating pad, and electric stove is to produce heat. However, in most electronic devices, the heat produced by resistors represents wasted power.

The power lost by the heating of resistors must be supplied by the power source. Since electrical power costs money, attempts are generally made to keep the power lost in resistors at a minimum.

Since resistors dissipate power, there must be some formula for power which involves resistance. Actually, there are two such formulas. One expresses power in terms of voltage and resistance. The other expresses power in terms of current and resistance. Let's see how these equations are derived.

First, let's consider how P can be expressed in terms of E and R. The basic formula for power is:

$$P = I \times E \text{ or } P = E \times I$$

However, from Ohm's law we know that

$$I = \frac{E}{R}$$

Thus, we can substitute

$$\frac{E}{R}$$

for I in the basic power equation. When we do this, the equation becomes:

$$P = E \times \frac{E}{R}$$

Or,

$$P = \frac{E^2}{R}$$

This formula is used when we wish to find the power but only the voltage and resistance are known. For example, how much power is dissipated by a 22 ohm resistor if the voltage drop across the resistor is 5 volts?

$$P = \frac{E^2}{R}$$

$$P = \frac{5^2}{22}$$

$$P = \frac{25}{22}$$

$$P = 1.136 \text{ watts}$$

114

How much power is delivered to a 16 ohm circuit by a 12 volt battery?

$$P = \frac{E^2}{R}$$

$$P = \frac{12^2}{16}$$

$$P = \frac{144}{16}$$

$$P = 9 \text{ watts}$$

In some cases, only the current and resistance will be known. By combining the basic power formula with one of the Ohm's law formulas, we can derive an equation in which P is expressed in terms of I and R. Recall that the basic power formula is:

$$P = I \times E$$

From Ohm's law we know that $E = IR$. Thus, we can substitute IR for E in the power formula. The equation becomes:

$$P = I \times IR$$

$$P = I^2 R$$

This is the formula we use when we wish to find power but only current and resistance are known. For example, how much power is dissipated by the circuit shown in Figure 4-7A?

$$P = I^2 R$$

$$P = 0.5^2 \times 40$$

$$P = 0.25 \times 40$$

$$P = 10 \text{ watts}$$

How much power is dissipated by R_1 in Figure 4-7B?

$$P = I^2 R_1$$

$$P = 2^2 \times 10$$

$$P = 4 \times 10$$

$$P = 40 \text{ watts}$$

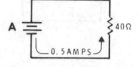

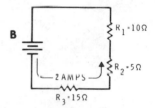

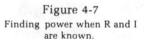

Figure 4-7
Finding power when R and I are known.

115

We can verify that the above equations are correct by working a sample problem three different ways. For example, consider the circuit shown in Figure 4-8. The current and voltage are given so we can use the formula P = IE:

$$P = IE$$

$$P = 0.5 \times 6$$

$$P = 3 \text{ watts}$$

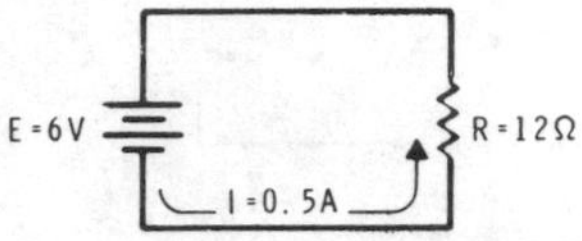

Or, since voltage and resistance are given, we can use the formula

$$P = \frac{E^2}{R}$$

Figure 4-8
Power can be found using either of the three power formulas.

$$P = \frac{E^2}{R} = \frac{6^2}{12} = \frac{36}{12} = 3 \text{ watts}$$

Or, since current and resistance are given, we can use the formula P = I²R:

$$P = I^2 R$$

$$P = 0.5^2 \times 12$$

$$P = 0.25 \times 12$$

$$P = 3 \text{ watts}$$

Notice that the same result is achieved regardless of the power equation used.

Deriving More Equations

Earlier we saw that the basic power equation (P = IE) can be rearranged to form current and voltage equations.

$$\text{Namely, } E = \frac{P}{I} \quad \text{and } I = \frac{P}{E} \text{ .}$$

In much the same way, the equation

$$P = \frac{E^2}{R}$$

can be rearranged to form an equation for voltage. That is,

$$P = \frac{E^2}{R}$$

116

Multiplying both sides by R we find:

$$P \times R = \frac{E^2}{R} \times R$$

The two R's on the right cancel:

$$P \times R = \frac{E^2}{\cancel{R}} \times \cancel{R}$$

Thus,

$$P \times R = E^2$$

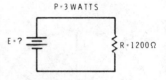

Now we can get rid of the exponent by taking the square root of both sides of the equation:

$$\sqrt{P \times R} = \sqrt{E^2}$$

Or,

$$\sqrt{P \times R} = E$$

Figure 4-9
Finding voltage when
resistance and power are known.

Notice that this equation expresses voltage in terms of power and resistance.

This equation allows us to work problems like the one shown in Figure 4-9. Here the resistance and power are given but the voltage is unknown:

$$E = \sqrt{PR}$$

$$E = \sqrt{3 \times 1200}$$

$$E = \sqrt{3600}$$

$$E = 60 \text{ volts}$$

NOTE:
Appendix A of this unit, explains how to find the square root of a number. If you are unfamiliar with this technique, read Appendix A at this time.

117

In the same way, the equation $P = I^2 R$ can be converted to a current equation. That is:

$$P = I^2 R$$

Dividing both sides by R and letting the R's on the right side cancel:

$$\frac{P}{R} = \frac{I^2 \cancel{R}}{\cancel{R}}$$

This leaves:

$$\frac{P}{R} = I^2$$

Taking the square root of both sides:

$$\sqrt{\frac{P}{R}} = I$$

Thus, current can be expressed in terms of power and resistance.

By rearranging the equations

$$P = I^2 R$$

and

$$P = \frac{E^2}{R}$$

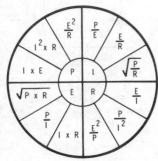

Figure 4-10
Wheel diagram for finding the proper equation for current voltage, resistance, and power.

equations for R can be obtained. The derivation is left to the reader but the final equations are:

$$R = \frac{P}{I^2} \quad \text{and } R = \frac{E^2}{P}$$

In this unit, we have discussed twelve important formulas. Figure 4-10 is a wheel diagram which may help you to remember these equations. The inner circle contains the four basic units: power, current, voltage, and resistance. The outer circle contains three formulas for each quantity. For example, the three formulas for resistance are:

$$R = \frac{E}{I}, \quad R = \frac{E^2}{P}, \text{ and } \quad R = \frac{P}{I^2}$$

Using this diagram, you can quickly find the proper formula for any problem in which two of the quantities are known.

While it is not necessary to memorize these 12 equations, you should memorize the basic Ohm's law and power formulas as soon as possible. That is, you should memorize these two equations:

$$E = I R \text{ and } P = I E$$

If you know these two formulas, you can derive any of the other formulas with some mathematical manipulations as you have seen. Until you become more familiar with these equations, use the diagram shown in Figure 4-10 as an aid.

SUMMARY

The following is a point by point summary of this unit.

Ohm's law defines the relationship between the three fundamental electrical quantities. It describes how current, voltage, and resistance are related.

Formulas are used to define the relationship. In these formulas, the letter I is used to represent current; E is used to represent voltage; and R is used to represent resistance.

The current formula states that current is equal to voltage divided by resistance. Or, stated as an equation:

$$I = \frac{E}{R}$$

The unit of current in the equation depends on the units of voltage and resistance. Assuming that emf is expressed in volts, the current will be in amperes if the resistance is in ohms. However, the current will be in milliamperes if the resistance is in kilohms. Finally, the current will be in microamperes if the resistance is in megohms.

The voltage formula states that voltage is equal to the current multiplied by the resistance. That is, $E = I \times R$.

E is expressed in volts when I is in amperes and R is in ohms. E is also in volts when I is in milliamperes and R is in kilohms. Finally, E is in volts when I is in microamperes and R is in megohms.

The resistance formula states that resistance is equal to voltage divided by current. That is,

$$R = \frac{E}{I}$$

Assuming that E is given in volts, R will be in ohms when I is in amperes. However, R will be in kilohms when I is in milliamperes. Finally, R will be in megohms when I is in microamperes.

While it is handy to remember which of the metric prefixes go together when solving Ohm's law problems, it is not absolutely essential. If you will convert all quantities to volts, ohms, and amperes, you can solve any Ohm's law problem without worrying about metric prefixes.

Power is defined as the rate at which work is done. The unit of work is the joule while the unit of power is the watt. The watt is equal to one joule per second.

Expressed in English units, the watt is approximately equal to three-fourths of a foot-pound of work per second. The horsepower is equal to 746 watts or about three-fourths of a kilowatt.

The watt can also be expressed in terms of current and voltage. The watt is the amount of power expended when one volt of emf causes one ampere of current.

The letter P is used to represent power in equations. Power is equal to voltage multiplied by current. Or, $P = I \times E$.

Power can also be expressed in terms of current and resistance or in terms of voltage and resistance. Figure 4-10 lists several important equations involving power, current, voltage, and resistance. You should study this list carefully.

Appendix A

FINDING THE SQUARE ROOT OF A NUMBER

The easiest way to find the square root of a number is to use a calculator which has a square root key. If your calculator does not have a square root key, you may still be able to compute square roots using a procedure described in the instruction manual supplied with your calculator. For those without a calculator, a slide rule is the next easiest method. Tables or logarithms may also be used. In spite of all these aids, you may on some occasion find it necessary to use the old-fashioned, long-hand method of computing the square root of a number. The following programmed instruction sequence explains this method of computing square roots.

---

1. The square root of a given number is the number which when multiplied by itself equals the given number. For example, the square root of 25 is 5 because $5 \times 5 = $ _____.

---

2. (25) The symbol which is used in mathematics to indicate that the square root is to be taken is the radical sign ($\sqrt{}$). Thus, $\sqrt{25} = $ _____.

---

3. (5) Some examples are so simple that we intuitively know the answer. For example, $\sqrt{9} = 3$, $\sqrt{100} = 10$, $\sqrt{64} = $ _____.

---

4. (8) However, when the number is very long such as 53,545.96 the square root is found by a long (but relatively simple) procedure. Let's find the square root of the number by using this procedure.

   The first step is to write the number in groups of two digits starting at the decimal point. For example, if we were taking the square root of 6,314.313 we would group the digits like this:

   $$63 \quad 14 \quad . \quad 31 \quad 30$$

   However, in our example the number is 53,545.96. Therefore, the number is written in groups of two digits like this: _____.

---

5. (5 35 45 . 96) Notice that 5 must be written by itself in this example because there are an odd number of digits to the left of the decimal point. Since we are taking the square root of this number, it should be placed under a radical sign like this: _____.

6. ($\sqrt{5}$ 35 45 . 96) We are now ready to find the first number in the answer. We do this by examining the first group of numbers on the left. In this case, the "group" consists of the single number 5. On the line above the 5, we place a number which when multiplied by itself will come close to equalling, but will not exceed 5. In this case we use 2 because $2 \times 2 = 4$. This is as close as we can get to 5 without exceeding 5. Thus, the first digit in the answer is _____.

---

7. (2) Our work to this point looks like this:

$$\begin{array}{r} 2 \qquad\qquad\quad \\ \hline \sqrt{5\ \ 35\ \ 45\ \ .\ \ 96} \end{array}$$

Next we multiply 2 by itself and place the product under 5 and subtract. Thus, our work becomes:

$$\begin{array}{r} 2 \qquad\qquad\quad \\ \hline \sqrt{5\ \ 35\ \ 45\ \ .\ \ 96} \\ 4 \qquad\qquad\qquad\quad \\ \hline 1 \qquad\qquad\qquad\quad \end{array}$$

At this point, the next group of two digits is brought down and placed beside the 1. The next two digits are _____.

---

8. (35) Thus, our work looks like this:

$$\begin{array}{r} 2 \qquad\qquad\quad \\ \hline \sqrt{5\ \ 35\ \ 45\ \ .\ \ 96} \\[4pt] 4 \qquad\qquad\qquad\quad \\ \hline 1\ \ 35 \qquad\qquad\qquad \end{array}$$

123

The next step is to double the number above the radical sign and place the result in front of the 135 like this:

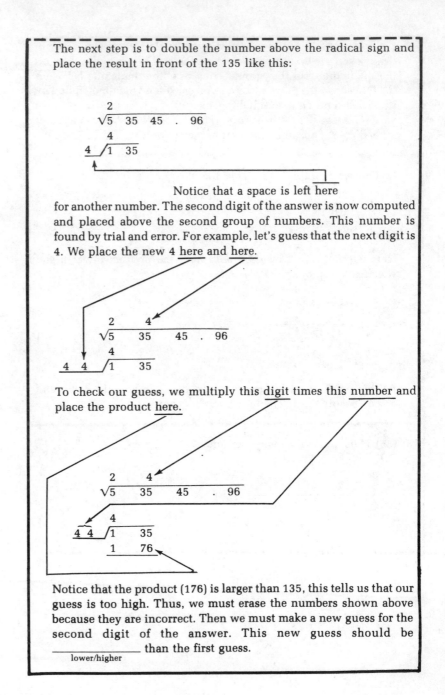

Notice that a space is left here for another number. The second digit of the answer is now computed and placed above the second group of numbers. This number is found by trial and error. For example, let's guess that the next digit is 4. We place the new 4 here and here.

To check our guess, we multiply this digit times this number and place the product here.

Notice that the product (176) is larger than 135, this tells us that our guess is too high. Thus, we must erase the numbers shown above because they are incorrect. Then we must make a new guess for the second digit of the answer. This new guess should be _____ than the first guess.

lower/higher

124

9. (lower) Let's try 3 as the second digit of the answer. Thus, our problem becomes:

```
        2     3
    √5   35    45    .    96

        4
  43  / 1    35
```

Now we multiply 3 times 43 and place the product under the 135. If the product is smaller than 135 we subtract:

```
        2     3
    √5   35    45    .    96

        4
  4 3  / 1    35
         1    29
                6
```

This number tells us that our guess is correct. If this number is greater than 43, our guess is too low. On the other hand we saw in Frame 8 what happens if our guess is too high. Thus, at this point, we know that the first two digits in the answer are _____ _____.

10. (23) In the next step we bring down the next two digit group of numbers so that our work looks like this:

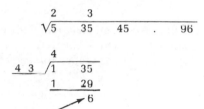

```
        2     3
    √5   35    45    .    96

        4
  4 3  / 1    35
         1    29
                6    45
```

125

As before the number above the radical sign (23) is doubled to 46. This 46 is placed in front of the 645. A space is left open after the 46 for the next trial answer:

```
          2    3
      √5   35   45    .    96

            4
     43  /1   35

          1   29
    46   /        6   45
      ↑
    space
```

We guess at the third digit in the answer and place our guess above the third group of numbers and in front of the 46 on the bottom line. By trial and error you will soon find that the proper number for this third digit is _____.

---

11. (1) Thus, the problem becomes:

```
          2      3      1
      √5    35     45    .    96

             4
    4 3  /1    35
          1    29
    4 6 1 /          6     45
```

Next, we multiply 1 × 461, place the product under 645, and subtract:

```
          2      3      1
      √5    35     45    .    96

             4
    4 3  /1    35

          1    29
    4 6 1 /          6     45
                     4     61
                     1     84
```

The next step is to bring down the final group of two numbers so that the number on the bottom line becomes _____.

12. (1 84 96) The decimal point in the answer is always immediately above the decimal point in the number under the radical sign. Thus, by placing the decimal point in the answer our work looks like this:

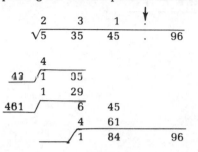

To find the next digit of the answer, we double our partial answer (231) to get 462. We place this number in front of the 1 84 96 leaving a space for our trial number:

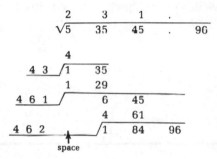

Again by trial and error, we find that the final digit must be
_____.

127

13. (4) Thus, we place 4 in the answer and after the 462. Next we multiply
    4 × 4624 so that our final work looks like this:

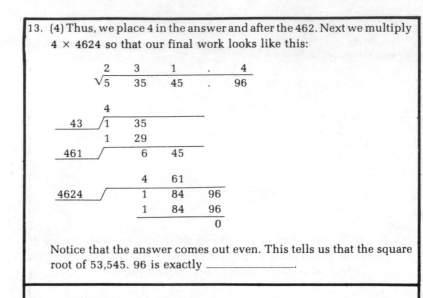

```
                2     3     1     .     4
            √5      35    45     .   96

                4
        43   /1     35
              1     29
       461   /        6    45

                4    61
      4624   /        1    84    96
                      1    84    96
                                  0
```

Notice that the answer comes out even. This tells us that the square
root of 53,545. 96 is exactly _____.

14. (231.4) We can easily prove this by multiplying 231.4 by itself.

Thus: 231.4
         × 231.4
         _____
            925 6
           2314
           6942
           4628
         _____
         53,545.9 6

128

To be sure you have the idea, an additional example is given below. Study each step carefully starting with Step 1 and ending with Step 23.

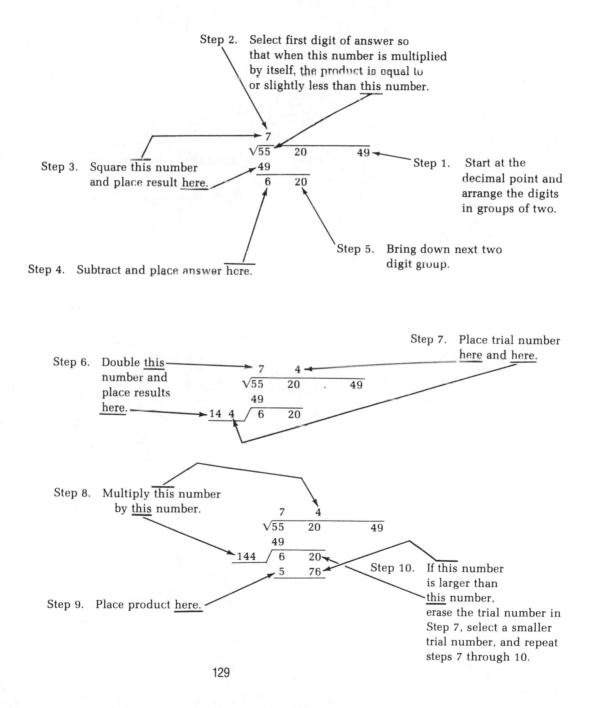

Step 2. Select first digit of answer so that when this number is multiplied by itself, the product is equal to or slightly less than this number.

Step 3. Square this number and place result here.

Step 1. Start at the decimal point and arrange the digits in groups of two.

Step 4. Subtract and place answer here.

Step 5. Bring down next two digit group.

Step 6. Double this number and place results here.

Step 7. Place trial number here and here.

Step 8. Multiply this number by this number.

Step 9. Place product here.

Step 10. If this number is larger than this number, erase the trial number in Step 7, select a smaller trial number, and repeat steps 7 through 10.

Step 14. Place decimal point in answer <u>here.</u>

Step 13. Bring down next two digit group.

Step 11. Subtract and place difference <u>here.</u>

Step 12. If difference is larger than <u>this</u> number, erase the trial number in Step 7, select a larger trial number and repeat steps 7 through 12.

Step 15. Double <u>this</u> number and place result <u>here.</u>

Step 16. Place trial number <u>here</u> and <u>here.</u>

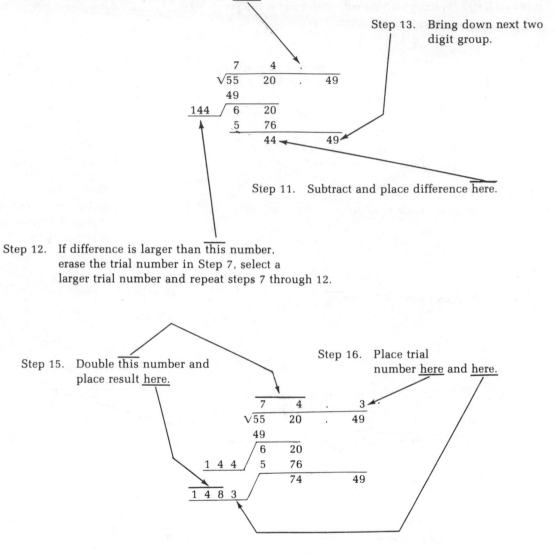

130

Step 17.   Multiply $\overline{\text{this}}$ number by <u>this</u> number.

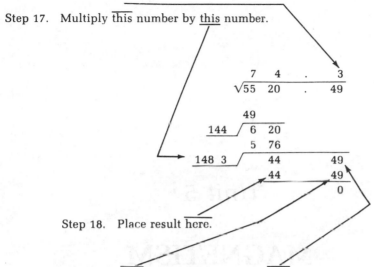

```
              7   4   .    3
            √55  20   .   49

              49
       144 /  6  20
              5  76
     148 3 /      44        49
                  44        49
                            0
```

Step 18.   Place result <u>here</u>.

Step 19.   If $\overline{\text{this}}$ number is larger than $\overline{\text{this}}$
           number, erase the trial number in Step 16, select a smaller trial
           number and repeat steps 16 through 19.

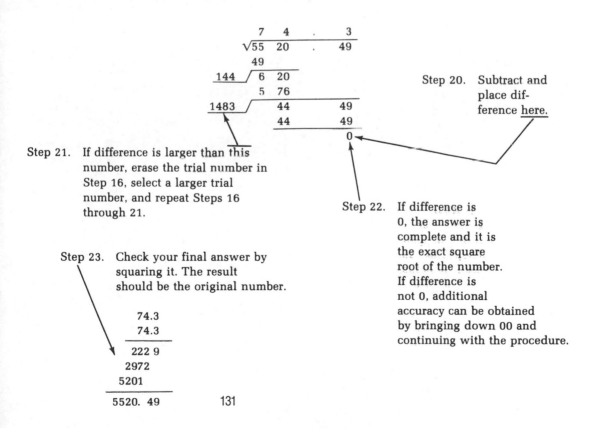

```
              7   4   .    3
            √55  20   .   49
              49
       144 /  6  20
              5  76
      1483 /     44        49
                 44        49
                           0
```

Step 20.   Subtract and
           place dif-
           ference <u>here</u>.

Step 21.   If difference is larger than $\overline{\text{this}}$
           number, erase the trial number in
           Step 16, select a larger trial
           number, and repeat Steps 16
           through 21.

Step 22.   If difference is
           0, the answer is
           complete and it is
           the exact square
           root of the number.
           If difference is
           not 0, additional
           accuracy can be obtained
           by bringing down 00 and
           continuing with the procedure.

Step 23.   Check your final answer by
           squaring it. The result
           should be the original number.

```
         74.3
         74.3
        ─────
         222 9
        2972
        5201
       ───────
       5520. 49
```

131

Unit 5

MAGNETISM

INTRODUCTION

In any study of electricity or electronics, the effects of magnetism must be considered. Electricity and magnetism are inseparable. Electric currents produce magnetic fields and in special cases magnetic fields produce electric currents. Electricity and magnetism are often considered to be two different aspects of a more general effect called *electromagnetic phenomenon*.

THE MAGNETIC FIELD

In science, action-at-a-distance is explained in terms of fields. For example, you have seen that a charged particle can attract or repel another charged particle simply by coming close to it. A region of electrical influence extends outside each particle. This region of influence is called a *field*. An electrical field made up of lines of force is said to exist around every charged particle.

The field concept is also used to explain why certain metals can attract other metals. Everyone knows that a magnet attracts small pieces of iron or steel. A region of influence extends outside the magnet into the surrounding space. In this case, the region is called a *magnetic* field and is said to be made up of *magnetic* lines of force. Thus, a magnet is a piece of material which has a magnetic field surrounding it.

Magnets

Magnets may be classified in several different ways. First, they can be classified according to the method by which they obtain their magnetic field. The first known magnets were *natural* magnets called magnetite or lodestone. These materials have a magnetic field when found in their natural state. *Artificial* magnets can be created from natural magnets. For example, if soft iron is rubbed repeatedly over a piece of lodestone, a magnetic field is transferred to the iron. Another type of magnet is the electromagnet. Its magnetic field is produced by an electric current. We will discuss the electromagnet in detail later.

Magnets are often classified by their shape. Thus, there are horseshoe magnets, bar magnets, ring magnets, etc.

Some materials readily retain their magnetic fields for long periods of time. These are called *permanent* magnets. Other materials quickly lose their magnetism and are called *temporary* magnets. Both of these types find wide use in electronics.

Finally, magnets are classified by the type of material used such as *metallic* magnets or *ceramic* magnets. Often this is carried even further and they are named according to the alloy used. Two popular classifications are Alnico (an alloy of aluminum, nickel, and cobalt) and Cunife (an alloy of copper, nickel, and iron).

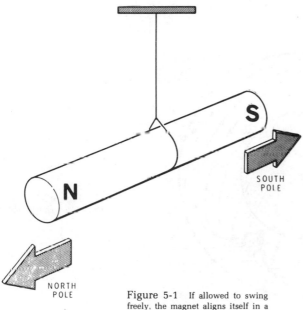

SOUTH
POLE

NORTH
POLE

Figure 5-1   If allowed to swing
freely, the magnet aligns itself in a
north-south direction.

The two ends of a magnet have different characteristics. One end is called a
south (S) pole while the other is called a *north (N)* pole. One reason for
choosing these names is that a bar magnet will align itself in a north-south
direction if allowed freedom of movement as shown in Figure 5-1. The
north (N) pole of the magnet is defined as that end which points toward the
north pole of the Earth.

The magnet lines up in this way because the Earth itself is a huge magnet.
As shown in Figure 5-2, it has its own magnetic field which influences any
magnet on Earth. This fact has been used for centuries by mariners and
explorers who rely on the magnetic compass. The compass itself is nothing
more than a tiny magnet balanced on a pin so that it rotates freely.

Magnets tend to align in a north-south direction because of a fundamental
law of magnetism. This law states that like poles repel while unlike poles
attract. Figure 5-3 illustrates this point. To see why magnets behave in this
way, we must consider the nature of the magnetic lines of force.

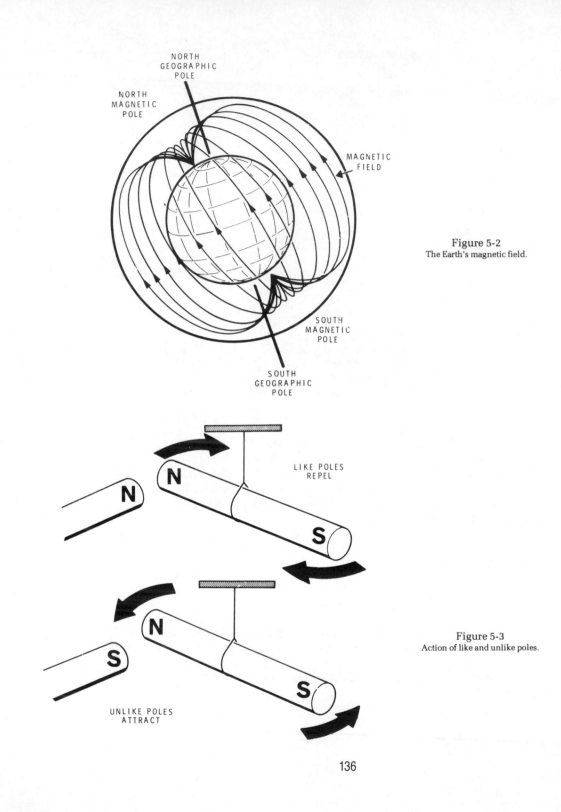

NORTH
GEOGRAPHIC
POLE

NORTH
MAGNETIC
POLE

MAGNETIC
FIELD

SOUTH
MAGNETIC
POLE

SOUTH
GEOGRAPHIC
POLE

Figure 5-2
The Earth's magnetic field.

LIKE POLES
REPEL

N

N

S

N

S

S

UNLIKE POLES
ATTRACT

Figure 5-3
Action of like and unlike poles.

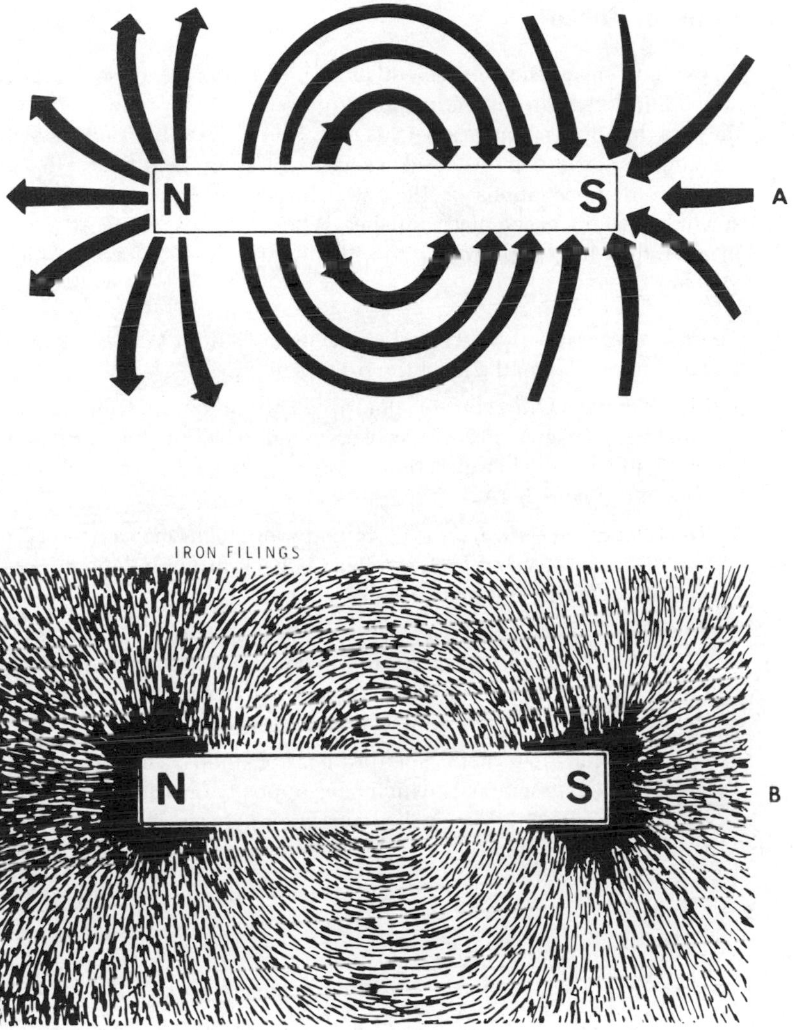

IRON FILINGS

Figure 5-4   Lines of force
surround the magnet.

Lines of Force

To explain a magnetic field, scientists assume that lines of magnetic force called *flux lines* surround a magnet. Figure 5-4A shows the flux lines as they might appear around a bar magnet. While these lines are invisible, their effects can be demonstrated in several different ways. One of the most dramatic demonstrations is illustrated in Figure 5-4B. Iron filings are sprinkled evenly over a piece of paper. When the paper is placed over the magnet, the filings align so that the effects of the lines of force are clearly visible.

There are several basic rules and characteristics of flux lines which you should know. Four of the most important are:

1. *Flux lines have direction or polarity.* The direction of the flux lines outside the magnet are arbitrarily assumed to be from the north pole to the south pole. This direction is often indicated by arrowheads as shown in Figure 5-4A.

2. *The lines of force always form complete loops.* This may not be obvious from Figure 5-4A, but each line curves back through the body of the magnet to form a complete loop.

3. *Flux lines cannot cross each other.* This is the reason that like poles repel. Lines which have the same polarity can neither connect nor cross. When one field intrudes into another as shown in Figure 5-5A, the lines repel and the magnets tend to move apart.

4. *Flux lines tend to form the smallest possible loops.* This explains why unlike poles attract. Lines which have opposite polarity can link up as shown in Figure 5-5B. Then the loops attempt to shorten by pulling the two magnets together.

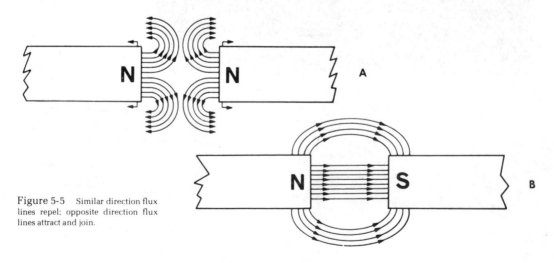

Figure 5-5   Similar direction flux lines repel; opposite direction flux lines attract and join.

138

Magnetic Materials

Of the 92 natural elements, only three respond readily to magnetic fields. These are iron, cobalt, and nickel. All three are metals and they have atomic numbers of 26, 27, and 28 respectively. Each has two valence electrons so their chemical and electrical characteristics are quite similar. In addition to these elements, there are dozens of alloys which have magnetic characteristics. Substances which readily respond to magnetic fields are called *ferromagnetic substances*. Ferromagnetic materials are strongly attracted by a magnetic field.

Most substances are classified as *paramagnetic*. These are substances which are attracted only slightly by a strong magnetic field. Generally, the force of attraction is so tiny, that these materials are considered to be non-magnetic. Substances such as air, aluminum, and wood are paramagnetic in nature.

Technically speaking, there is one other classification called *diamagnetic*. Diamagnetic materials are slightly repelled by magnetic fields. However, here again, the force of repulsion is so tiny that these materials are generally considered non-magnetic. Examples of diamagnetic materials are bismuth, quartz, water, and copper.

The characteristic which determines if a substance is ferromagnetic, paramagnetic, or diamagnetic is called *permeance* or *permeability*. Permeability refers to the ability of various materials to accept or allow magnetic lines of force to exist in them. Air is considered the standard with a permeability of 1. Other substances are given a permeability rating lower or higher than one depending on their magnetic characteristics. Iron is about 7000 times more effective in accepting flux lines than is air. Consequently, iron has a permeability of about 7000. Figure 5-6 is a table showing the relative permeabilities of several different substances. Notice that those substances which have values of permeability less than 1 are diamagnetic and are slightly repelled by flux lines. Those having values slightly greater than 1 are paramagnetic and are slightly attracted by flux lines. Finally, those having permeabilities much higher than 1 are ferromagnetic and are strongly attracted by flux lines.

Permeability of materials can be compared to the conductance in a circuit. Recall that conductance indicates the ease with which a material or circuit will allow current to flow. In much the same way, permeability is the ease with which a material will accept lines of flux.

139

MATERIAL	PERMEABILITY	CHARACTERISTIC	ACTION
BISMUTH	0.999833	DIAMAGNETIC	SLIGHTLY REPELLED
WATER	0.999991	DIAMAGNETIC	SLIGHTLY REPELLED
COPPER	0.999995	DIAMAGNETIC	SLIGHTLY REPELLED
AIR	1.000000	PARAMAGNETIC	NON-MAGNETIC
OXYGEN	1.000002	PARAMAGNETIC	SLIGHTLY ATTRACTED
ALUMINUM	1.000021	PARAMAGNETIC	SLIGHTLY ATTRACTED
COBALT	170.	FERROMAGNETIC	STRONGLY ATTRACTED
NICKEL	1000.	FERROMAGNETIC	STRONGLY ATTRACTED
IRON	7000.	FERROMAGNETIC	STRONGLY ATTRACTED
PERMALLOY*	100,000.	FERROMAGNETIC	STRONGLY ATTRACTED
SUPERMALLOY**	1,000,000.	FERROMAGNETIC	STRONGLY ATTRACTED

*PERMALLOY-AN ALLOY OF 17% IRON, 4% MOLYBDENUM, 79% NICKEL
**SUPERMALLOY-AN ALLOY OF 16% IRON, 5% MOLYBDENUM, 79% NICKEL

Figure 5-6   Relative
permeabilities of materials.

Now let's see why permeability is so important. Figure 5-7A shows a permanent magnet surrounded by its lines of flux. Figure 5-7B shows how the flux lines are distorted when a piece of iron is brought near the magnet. Since iron has a high permeability it can support lines of flux much more easily than the surrounding air. Consequently, a large number of the flux lines pass through the iron bar. At the same time the lines attempt to contract to the smallest possible loops. This tends to attract the iron bar to the magnet.

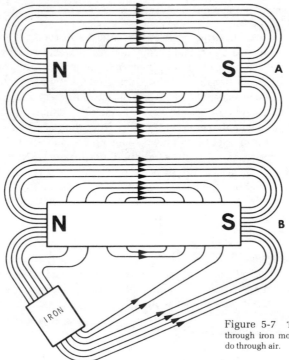

Figure 5-7   The flux lines pass through iron more easily than they do through air.

140

Theory of Magnetism

While it is difficult to explain exactly what magnetism is, a theory has been developed which explains the observed phenomenon. As with the basic theories of electricity, this one starts with the electron. We have seen that the electron orbits the nucleus of the atom in much the same way that the Earth orbits the sun. It also appears that the electron spins on its axis as shown in Figure 5-8A in much the same way that the Earth does. We have also seen that the electron has an electrostatic field as shown in Figure 5-8B. It appears to be a fact of nature that a moving electrical charge produces a magnetic field. The magnetic field produced by the spinning charge exists as concentric circles around the electron as shown in Figure 5-8C. The direction of the magnetic field depends on the direction of spin of the electron. At any given point, the electrostatic field is at right angles to the magnetic field. These combined fields at right angles are often called an *electromagnetic field*. Figure 5-8D shows the complete picture of the electron.

We have seen that iron, nickel, and cobalt are the only natural magnetic elements. Each of these elements has two valence electrons. In other substances, the electrons tend to pair off with electrons of *opposite* spin. This means that the electrons have opposite magnetic characteristics which tend to cancel. However, in iron, nickel, and cobalt the two valence electrons have the same direction of spin. Consequently, their magnetic fields do not cancel; they add. Thus, an atom of iron, nickel, or cobalt has a net magnetic field.

Small groups of these atoms tend to form tiny permanent magnets called magnetic *domains*. When not in the presence of a magnetic field, these domains are arranged haphazardly as shown in Figure 5-9A. Because, the domains are turned at odd angles, the net magnetic effect is zero. A piece of metal such as this can be magnetized by subjecting it to a strong magnetic field. As shown in Figure 5-9B this causes all the domains to align in the same direction. With all the domains aligned in a common direction, the entire piece of metal becomes a magnet.

There are several experiments which seem to verify the domain theory. The first is shown in Figure 5-10A. If a bar magnet is cut into several pieces, each piece becomes a complete magnet having both a north and a south pole. Figure 5-10B shows another experiment. When the magnet is hit with the hammer, the domains are jarred back to a random pattern and the net magnetism is lost. Figure 5-10C shows that the same thing happens when the magnet is heated. The heat energy causes the domains to vibrate enough to rearrange themselves in a random pattern.

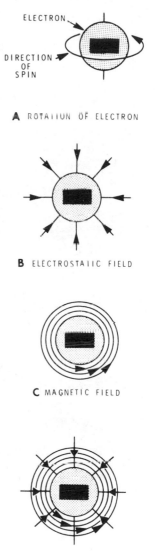

A ROTATION OF ELECTRON

B ELECTROSTATIC FIELD

C MAGNETIC FIELD

D ELECTROMAGNETIC FIELD

Figure 5-8
The electron's role in magnetism.

141

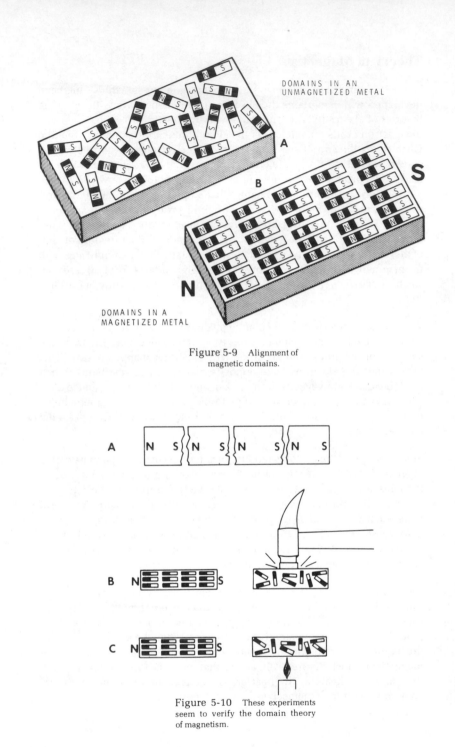

DOMAINS IN AN
UNMAGNETIZED METAL

A

B

S

N

DOMAINS IN A
MAGNETIZED METAL

Figure 5-9   Alignment of
magnetic domains.

Figure 5-10   These experiments
seem to verify the domain theory
of magnetism.

ELECTRICITY AND MAGNETISM

Electricity and magnetism are closely related. The electron has both an electrostatic field and a magnetic field. This may lead to the conclusion that a charged object should have a magnetic field. However, this is not the case, since the magnetic field of about half the electrons will be opposite that of the other half. Nevertheless, the electron plays an important part in magnetism. It can be forced to produce a magnetic field in substances which are normally considered non-magnetic such as copper and aluminum. The key is *motion*. Motion is the catalyst which links electricity and magnetism. Anytime a charged particle moves, a magnetic field is produced. If a large number of charged particles can be moved in a systematic way, a usable magnetic field is formed. You have seen that current flow is the systematic movement of large numbers of electrons. Thus, current flow causes a magnetic field.

Current Flow and Magnetism

When current flows through a wire; a magnetic field is developed around the wire. The field exists as concentric circles as shown in Figure 5-11. While this field has no north or south pole, it does have direction. The direction of the field depends on the direction of current flow. The arrow heads on the flux lines indicate their direction. This does not mean that the flux lines are moving in this direction. It simply means that they are *pointing* in this direction.

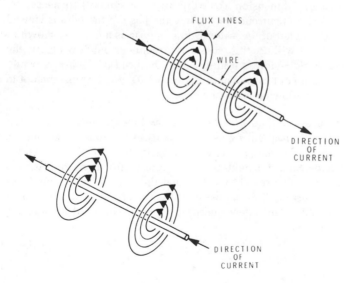

Figure 5-11 Flux lines exist as concentric circles around a current carrying conductor.

143

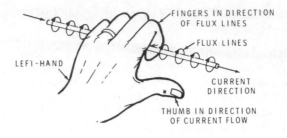

Figure 5-12
Left-hand magnetic field rule.

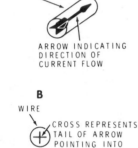

A

WIRE

ARROW INDICATING
DIRECTION OF
CURRENT FLOW

B

WIRE

CROSS REPRESENTS
TAIL OF ARROW
POINTING INTO
THE PAGE

C

D

DOT REPRESENTS
HEAD OF ARROW
POINTING OUT
OF PAGE

Figure 5-13 Developing two
new symbols to represent current
flow in a third dimension.

The direction of the flux lines can be determined if the direction of the current flow is known. The rule for determining this is called *the left-hand magnetic-field rule* or *the left-hand rule for conductors*. It is illustrated in Figure 5-12. This rule states:

If you grasp the conductor in your *left* hand with your thumb pointing in the direction of current electron flow through the conductor, your fingers now point in the direction of the flux lines.

Study Figure 5-12 until you understand this rule. Try this rule on the two conductors shown in Figure 5-11.

In explaining some aspects of electromagnetism, it is helpful to show current flow in a third dimension. To do this, two new symbols are necessary. Figure 5-13A shows current flowing into the page. If the wire is viewed from the end, the tail of the arrow would appear as a cross as shown in Figure 5-13B. We will use this cross to represent current flowing into the page. If this wire is viewed from the end, the head of the arrow would appear as a round dot as shown in Figure 5-13D. We will use the dot to represent current flowing out of the page.

Figure 5-14 uses these new symbols to show how opposite and similar currents establish magnetic fields. In Figure 5-14A, opposite currents are shown. Using the left-hand rule, you can verify the direction of the two magnetic fields. Since the fields point in opposite directions, they tend to repel each other. Figure 5-14B shows that the opposite situation exists when the two currents flow in the same direction. Here the fields point in the same direction. Thus, they are free to connect. This tends to draw the two fields together.

144

As long as the conductor is a straight piece of wire, the magnetic field produced is of little practical use. Although it has direction, it has no north or south pole. Also, unless the current is extremely high, the magnetic field has little strength. However, by changing the shape of the wire, we can greatly improve its magnetic characteristics.

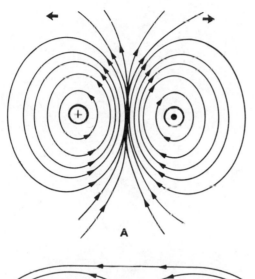

A

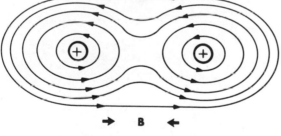

B

Figure 5-14 Opposite currents cause opposite fields which repel; Currents in the same direction cause fields which add and attract.

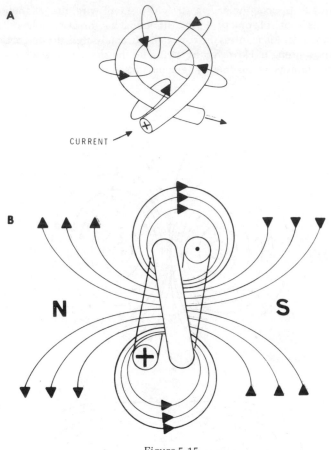

Figure 5-15
Flux lines around a loop of wire.

Figure 5-15 shows two views of a short piece of wire twisted into a loop. Simply forming the loop helps the magnetic characteristics in three ways. First, it brings the flux lines closer together. Second it concentrates the majority of the flux lines in the center or *core* of the loop. Third, it creates north and south poles. The north pole is the side where the flux lines come out; the south pole, the side where they go in. Thus, this loop of wire has the characteristics of a magnet. In fact, this is an example of a simple electromagnet.

The Electromagnet

Electromagnetism is used in many different electronic devices. In its simplest form, the electromagnet is nothing more than a length of wire wrapped in coils as shown in Figure 5-16. When current passes through the wire, a magnetic field is established. Because the turns of wire are very close together, the flux lines of the individual turns add together, to produce a very strong magnetic field. The more turns in the coil, the more flux lines there will be to add together. Also, the more current that flows through the coil, the more flux lines there will be. Consequently, the strength of the magnetic field is directly proportional to both the number of turns in the coil and the amount of current through the coil.

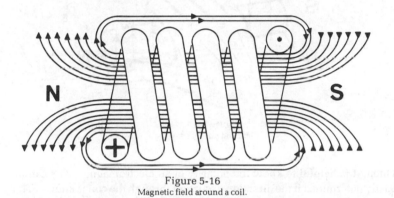

Figure 5-16
Magnetic field around a coil.

The magnetic field around the coil has the same characteristics of the magnetic field around a permanent magnet. However, one difference is that the field around the coil exists only when current flows through the coil. Another important difference is that the strength of the magnetic field around the coil can be varied by changing the amount of current flowing through the coil.

We have seen two ways to increase the strength of the magnetic field around an electromagnet. One way is to increase the current. Another is to increase the number of turns. However, a third method is the most dramatic of all. It involves the addition of a bar of ferromagnetic material, called a *core*, to the center of the coil. For example, if an iron core is slipped into the coil shown in Figure 5-16, the strength of the magnetic field would greatly increase. The reason for this is that the iron core has a much higher value of permeability than air. Consequently, the iron core can support many times more flux lines than air. Most electromagnets are made by winding many turns of wire around a bar of ferromagnetic material such as iron.

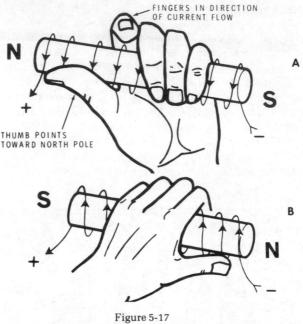

FINGERS IN DIRECTION
OF CURRENT FLOW

N

A

+

S

–

THUMB POINTS
TOWARD NORTH POLE

S

B

+

N

–

Figure 5-17
Left-hand rule for coils.

Often, it is helpful to know the polarity of an electromagnet. This can be easily determined if the direction of current through the coil is known. The rule is called the *left-hand rule for coils* and is illustrated in Figure 5-17. The rule states:

   If you grasp the coil with your left hand in such a way that your fingers are wrapped around it in the same direction that current is flowing; your thumb will then point toward the north pole of the magnet.

Remember that current flows from negative to positive. In Figure 5-17A, the current flows up the back side of the coil and down the front side. If the fingers are wrapped in this direction, the thumb points to the left. Thus, the left end of the electromagnet is the North Pole. In Figure 5-17B, the coil is wrapped in the opposite direction and the north pole is on the right.

Using this same procedure, the direction of current flow can be determined if the north pole is known. Assume that we know the north pole is on the right but we do not know the direction of current. We simply grasp the coil in the left hand with the thumb pointing toward the north pole of the coil. The fingers now point in the direction of current flow.

148

Magnetic Quantities

In our study of electricity, we used electrical quantities such as voltage, current, resistance, conductance, and power. In much the same way, the study of magnetism requires that we learn several magnetic quantities. Of particular importance are the magnetic quantities: flux, flux density, magnetomotive force, field intensity, reluctance, and permeability. While the definitions of these quantities are straight forward, the units of measure for these quantities often become confusing The reason for this is that three different systems of measurements have been used over the years. The first is the English system which uses the familiar inches, pounds, etc. The two other systems are based on metric units. One is called the cgs system. Cgs stands for centimeter, gram, and second. Another system based on metric units is the mks system. Mks stands for meter, kilogram, and second. In the following discussion, the English units will be used because this is probably the system with which you are most familiar. At the end of this section, the English units are compared with the metric units.

Flux. The complete magnetic field of a coil or a magnet is known as the flux. Thus, the flux is the total lines of magnetic force. The Greek letter phi (Φ) is used to denote flux. In the English system, flux is measured in lines. A coil or magnet which produces 1000 lines of force has a flux of 1000 lines or 1 kiloline ($\Phi = 1$ K lines).

Flux Density. As the name implies, flux density refers to the number of lines per unit of area. In the English system the unit of area is the square inch. Thus, flux density is expressed as the number of lines per square inch. The letter B is used to represent flux density. If a coil with a cross sectional area of two square inches has a flux of 1000 lines then the flux density is 1000/2 or 500 lines per square inch (B = 500 lines/in^2).

Magnetomotive Force (mmf). Magnetomotive force is the force which produces the flux in an electromagnet or coil. As we have seen this force is directly proportional to the number of turns in the coil and the amount of current flowing through the coil. For this reason, the unit of mmf is the ampere-turn. This is the amount of force developed by one turn of wire when the current is one ampere. For example, a coil having 50 turns and a current of 2 amperes has a magnetomotive force of 50 x 2 = 100 ampere turns.

Field Intensity or Magnetizing Force. While mmf is a useful term, it is limited in application because it does not take into consideration the length of the coil. Thus, a coil with 50 turns may be 1 inch long or 10 inches long and still have the same mmf. Obviously though, the magnetic field would be concentrated in a much smaller space with the shorter coil.

Field intensity takes into consideration not only the mmf but also the length of the coil. Field intensity is expressed as ampere turns per inch and is represented by the letter H. For example, if a 2-inch coil has an mmf of 100 ampere turns, then the field intensity is 100/2 or 50 ampere turns per inch (H = 50 amp-turns/in). Field intensity is sometimes called *magnetizing force* which should not be confused with magnetomotive force.

Permeability. We have already discussed this important characteristic. You will recall that permeability is the ease with which a material can accept lines of force. It can also be thought of as the ability of a material to concentrate a large number of force lines in a small area. For example, a 1-inch column of soft iron can hold hundreds of times more flux lines than a comparable column of aluminum. The Greek letter mu (μ) is used to represent permeability.

Reluctance. The opposite or reciprocal of permeability is called *reluctance* and is represented by the letter R. Reluctance is generally defined as an opposition to flux. Thus, a material with high reluctance is reluctant to accept flux lines. Since reluctance is the reciprocal of permeability, it may be expressed by the equation:

$$R = \frac{1}{\mu}$$

For example, soft iron has a permeability of 2700. Thus, it has a reluctance of 1/2700. Since air has a permeability of 1, it has a reluctance of 1/1 or 1. Flux lines tend to follow the path of least reluctance.

Ohm's Law for Magnetic Quantities

Three of the above quantities are related by an equation which is very similar to the Ohm's law equation. In fact, these three magnetic quantities can be compared to the electrical units of current, voltage, and resistance. In this analogy, current corresponds to magnetic flux (Φ). You will recall that flux is produced by magnetomotive force (mmf). Consequently, mmf corresponds to voltage. Finally, the opposition to flux is called reluctance (R). Flux, mmf, and reluctance are related by the equation.

$$\Phi = \frac{\text{mmf}}{R}$$

This states that the magnetic flux developed in a core material is directly proportional to the magnetomotive force and inversely proportional to the reluctance. Because of the similarity of this equation to the Ohm's law equation discussed earlier, this is often called the Ohm's law for magnetic circuits.

Comparison of Units

As we have seen there are three systems of measurements in common use. The most familiar is the English system. However, the metric system is becoming increasingly popular. Unfortunately, the units for the various magnetic quantities are different in each system. The table shown in Figure 5-18 compares the various units.

In the English system, flux is measured in lines or kilolines; flux density is measured in kilolines per square inch; mmf is measured in ampere-turns; and field intensity is measured in ampere-turns per inch.

TERM	DESCRIPTION	SYMBOL	ENGLISH UNITS	CGS UNITS	MKS UNITS
FLUX	TOTAL LINES OF FORCE	$\emptyset$	LINES OR KILOLINES	MAXWELL 1 MAXWELL =1 LINE	WEBER 1 WEBER = 100,000,000 LINES
FLUX DENSITY	LINES PER UNIT AREA	B	$\dfrac{\text{KILOLINES}}{\text{IN}^2}$	GAUSS 1 GAUSS = $\dfrac{1 \text{ MAXWELL}}{\text{CM}^2}$	$\dfrac{\text{WEBER}}{\text{M}^2}$
MAGNETOMOTIVE FORCE	TOTAL FORCE THAT PRODUCES FLUX	MMF	$\dfrac{\text{AMPERE-}}{\text{TURN}}$	GILBERT 1 GILBERT = 0.796 AMP TURN 1 AMP-TURN = 1.25 GILBERT	AMPERE-TURN
FIELD INTENSITY OR MAGNETIZING FORCE	FORCE PER UNIT LENGTH OF FLUX PATH	H	$\dfrac{\text{AMPERE-}}{\text{TURN}}$ $\overline{\text{IN}}$	OERSTED 1 OERSTED = $\dfrac{1 \text{ GILBERT}}{\text{CM}}$	$\dfrac{\text{AMPERE-TURN}}{\text{M}}$

Figure 5-18
Comparison of magnetic units.

In the cgs system, four new terms (maxwell, gauss, gilbert, and oersted) are introduced. The maxwell is equal to one line of force. The gauss is equal to 1 maxwell per square centimeter. The gilbert is equivalent to about 0.8 ampere-turns. Or stated another way, the ampere turn is equal to about 1.25 gilberts. Finally, the oersted is equal to 1 gilbert per centimeter.

In the mks system, the unit of flux is the weber. The weber is equivalent to 10^8 or 100,000,000 lines of force. Flux density is expressed as webers per square meter. As with the English system, the mks unit of mmf is the ampere-turn. Finally, field intensity is expressed as ampere-turns per meter.

This table ignores permeability and reluctance because neither has a unit of measurement assigned to it in any of the three systems.

151

INDUCTION

Induction may be defined as the effect of one body on another without any physical contact between them. For example, in an earlier unit, you saw that a charged body can *induce* a charge in another body simply by coming close to it. That is, a charge can be induced into a body without physical contact. This is possible because an electrostatic field surrounds every charged body. Thus, the field of a charged body can affect another body without the two bodies actually touching. This is an example of electrostatic induction.

Magnetic Induction

Another type of induction is called *magnetic induction*. Everyone knows that a magnet can affect objects at a distance. A strong magnet can cause a compass needle to deflect over a distance of several feet. Another power of the magnet is to induce a magnetic field in a previously unmagnetized object. For example, a magnet can induce a piece of iron to become a magnet.

Figure 5-19 shows a bar of soft iron close to a permanent magnet. Notice that part of the field of the magnet passes through the iron bar. Magnetic lines of force enter the left side of the iron and exit on the right. This magnetic field causes the magnetic domains in the iron to line up in one direction. Thus, the piece of iron itself becomes a magnet. The south pole must be on the left since this is the end where the flux lines enter the iron. The north pole is on the right since the flux lines exit at this point. Notice that the north pole of the permanent magnet is closest to the induced south pole of the iron bar. Because these opposite poles attract, the iron bar is attracted to the magnet. Therefore, the attraction of a piece of iron by a permanent magnet is a natural result of magnetic induction.

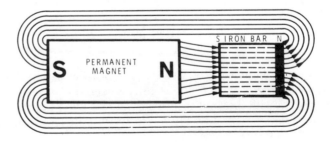

Figure 5-19   Magnetic induction.

152

When the piece of iron is removed from the magnetic field most of the magnetic domains return to random positions. However, a few of the domains will remain aligned in the north-south direction shown in Figure 5-19. Thus, the iron bar will retain a weak magnetic field even after it is removed from the influence of the permanent magnet. The magnetic field which remains in the iron bar is called *residual magnetism*. The ability of material to retain a magnetic field even after the magnetizing force has been removed is called *retentivity*. Soft iron has a relatively low value of retentivity. Thus, it retains little residual magnetism. Steel has a somewhat higher value of retentivity. Therefore, its residual magnetism is also higher. Some materials such as alnico have a very high value of retentivity. In these materials, the residual magnetic field is almost as strong as the original magnetizing field.

Electromagnetic Induction

Electromagnetic induction is the action that causes electrons to flow in a conductor when the conductor moves across a magnetic field. Figure 5-20 illustrates this action. When the conductor is moved up through the magnetic field, the free electrons are pushed to the right end of the conductor. This causes an excess of electrons at the right end of the conductor and a deficiency of electrons at the other end. The result is that a potential difference is developed between the two ends of the conductor. However, this potential difference exists only while the conductor is cutting the flux lines of the magnet. When the conductor moves out of the magnetic field, the electrons return to their original positions and the potential difference disappears. The potential difference also disappears if the conductor is stopped in the magnetic field. Thus, the conductor must move with respect to the flux lines before a potential difference is developed.

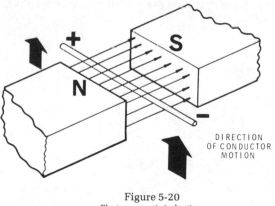

DIRECTION
OF CONDUCTOR
MOTION

Figure 5-20
Electromagnetic induction.

153

Motion is essential to electromagnetic induction. Some outside force must be applied to cause the conductor to move through the magnetic field. This mechanical force is converted to an electromotive force (emf) by electromagnetic induction. We say that an emf is induced into the conductor. The potential difference across the conductor is called an *induced emf* or an *induced voltage*.

The amount of emf induced into the conductor is determined by four factors:

1. The strength of the magnetic field.
2. The speed of the conductor with respect to the field.
3. The angle at which the conductor cuts the field.
4. The length of the conductor in the field.

The stronger the magnetic field, the greater the induced emf will be. Also, the faster the conductor moves with respect to the field, the greater the induced voltage will be. Relative motion between the conductor and the field can be produced by moving the conductor, by moving the field, or by moving both. The angle at which the conductor cuts the field is also important. Maximum voltage is induced when the conductor moves at right angles to the field as shown in Figure 5-20. Less voltage is induced when the angle between the lines of flux and the direction of motion of the conductor is less than 90°. In fact, if the conductor is moved parallel to the lines of flux as shown in Figure 5-21, no emf is induced at all. The fourth factor is the length of the conductor in the field. The longer the conductor, the greater the induced emf will be.

All four of these factors are a natural consequence of a basic law of electromagnetic induction. This law is called Faraday's law and it states:

The voltage induced in the conductor is directly proportional to the rate at which the conductor cuts the magnetic lines of force.

In other words, the more flux lines per second which are cut, the higher the induced emf will be.

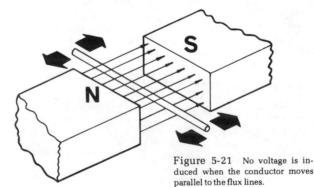

Figure 5-21 No voltage is induced when the conductor moves parallel to the flux lines.

The polarity of the induced emf can be determined by another of the left-hand rules. This one is called the *left-hand rule for generators* and is illustrated in Figure 5-22. It involves the thumb and the first two fingers of the left hand. The thumb is pointed in the same direction that the conductor is moving. The index or forefinger is pointed in the same direction as the flux lines. Now, the middle finger is pointed straight out from the palm at a right angle to the index finger. The middle finger is now pointing to the negative end of the conductor. This is the direction in which current will flow if an external circuit is connected across the two ends of the conductor.

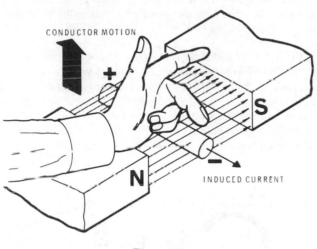

Figure 5-22
Left-hand rule for generators.

The AC Generator

Electromagnetic induction is important because it supplies virtually all of the electrical power used in the world today. It is the most efficient way known of producing electricity. Figure 5-23 shows a very basic electric generator. This device converts mechanical energy into electric energy by using electromagnetic induction. Mechanical energy is required to establish relative motion between the magnetic field and the conductor. Either the magnet or the conductor can be rotated. For this explanation, we will assume the conductor rotates in the counterclockwise direction. Notice that the conductor is shaped like a loop and is called an armature. When the loop or armature is rotated, one half moves up through the field near the south pole while the other half moves down through the field near the north pole.

If we apply the left hand generator rule to the side of the loop nearest the south pole, we find that the polarity of the induced voltage is negative at point A and positive at point B. Applying the same rule to the conductor near the north pole, we find that the induced voltage is negative at point C and positive at point D. Notice that the two induced voltages are series aiding. A meter connected across points E and F will measure the sum of the two induced voltages.

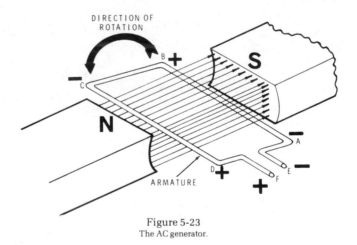

Figure 5-23
The AC generator.

Figure 5-24A shows the nature of the voltage which is induced into the armature. To see how this voltage is produced, we must follow the armature through one complete revolution. In each case, we will consider the voltage at point A with respect to point B. Figure 5-24B shows the armature at 90° increments. At 0°, the sides of the armature are moving parallel to the lines of flux. Thus, there is no induced voltage at this time. However, as the armature rotates, the loop begins cutting the lines of flux and a voltage is induced. Point A becomes positive with respect to point B. The voltage begins to rise and it reaches its maximum value at 90° of rotation. The voltage is maximum at this point because the armature is cutting the flux at a right angle. Thus, it cuts the maximum number of lines at this point. Once past 90°, the voltage begins to decrease because fewer lines per second are being cut. At 180°, the induced voltage is again zero because the armature is moving parallel to the lines. As the loop passes 180° and starts cutting the lines again, a voltage is once more induced. However, this time point A becomes negative with respect to point B. You can prove this by applying the left-hand generator rule. The maximum negative voltage is produced at 270° when the armature is once again cutting the lines at right angles. As the armature heads back toward its starting point the voltage begins to decrease back toward 0. At 360°, the armature is back where it started and the induced voltage is again zero.

The voltage shown in Figure 5-24A is called a *sine wave*. One cycle of the sine wave is produced for each revolution of the armature. If a load is connected across the armature between points A and B, current will flow through the load. For the first half cycle, current will flow from point B through the load to point A. However, during the next half cycle, current will flow from point A, through the load in the opposite direction to point B. Thus, during each cycle, the current reverses direction — flowing in one direction half of the time and flowing in the opposite direction the other half of the time. This is called alternating current and is abbreviated AC.

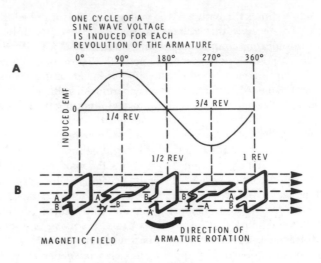

Figure 5-24
Generating a sine wave.

The voltages supplied to our homes, offices, and factories are ac voltages. The armature of the generator at the plant which provides our power is rotating 60 times each second. Thus, the voltage supplied by these power stations goes through 60 cycles like the one shown in Figure 5-24A each second. Most of the appliances in your home require 115 volts at 60 cycles.

The AC generator is often called an *alternator* because it produces alternating current. The simple machine shown here would not produce useful power because the armature consists of only one turn of wire. In a practical alternator hundreds of turns are wound into an armature which can produce considerable power.

The DC Generator

The AC generator or alternator can be converted to a DC generator. A device called a *commutator* is used to convert the AC voltage produced by the rotating loop into a DC voltage. Figure 5-25A shows how the commutator connects to the loop. The commutator is a cylinder shaped conductor. Two insulators are used to separate one half of the cylinder from the other half. Opposite sides of the loop are permanently connected to the opposite sides of the commutator. Thus, the commutator rotates with the loop. *Brushes* are used to make contact with the rotating commutator. The brushes are stationary and rest against opposite sides of the commutator. The brushes are made of a conducting material so that the emf produced by the loop is transferred to the brushes. In turn, wires are connected to the brushes so that the emf can be transferred to an external circuit.

The complete DC generator is shown in Figure 5-25B. Notice that it has four basic parts: a magnet to produce the magnetic field; a loop which produces the emf; a commutator which converts the induced emf to a DC voltage; and the brushes which transfer the DC voltage to an external circuit.

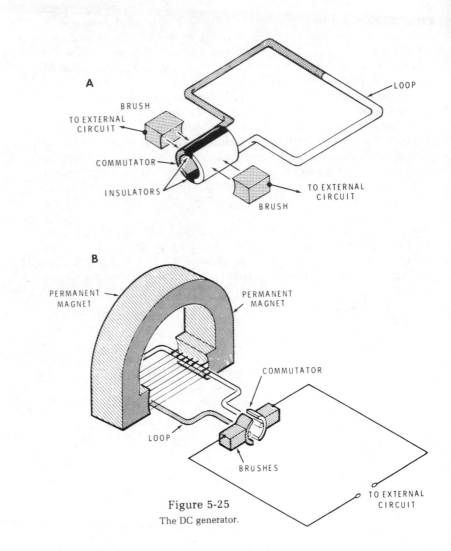

A

LOOP

BRUSH

TO EXTERNAL
CIRCUIT

COMMUTATOR

INSULATORS

TO EXTERNAL
CIRCUIT

BRUSH

B

PERMANENT
MAGNET

PERMANENT
MAGNET

COMMUTATOR

LOOP

BRUSHES

TO EXTERNAL
CIRCUIT

Figure 5-25
The DC generator.

Figure 5-26 illustrates the operation of the DC generator. At 0°, 180° and 360°, the sides of the loop are moving parallel to the flux lines and 0 volts is produced. This same situation existed with the AC generator. At 90° and 270°, the sides of the loop are cutting the flux lines at right angles. Thus, maximum voltage is produced. However, unlike the alternator, the voltage is positive at point A with respect to point B at both 90° and 270°. Let's see why.

Since the brushes are stationary and the commutator is rotating, each brush is alternately connected to opposite sides of the loop. When the magnetic field is in the direction shown, the side of the loop which is moving up through the field produces a negative voltage at the commutator. Also, the side of the loop moving down through the field produces a positive voltage at the commutator. Notice that the brush on the right is always connected to the side of the loop which is moving up through the field. Consequently, this brush is negative. On the other hand, the brush on the left is always connected to the side of the loop which moves down through the field. Consequently, this brush is positive. Thus if an external circuit is connected across A and B, current will always flow in the same direction i.e., from B to A.

The nature of the induced emf is shown in Figure 5-26B. This is called a pulsating DC voltage — DC because the current always flows in the same direction and pulsating because the level fluctuates. A pulsating DC voltage like this one is of little use in this form. However, as you will see in a future unit, this type of voltage can be smoothed out to form a constant DC voltage.

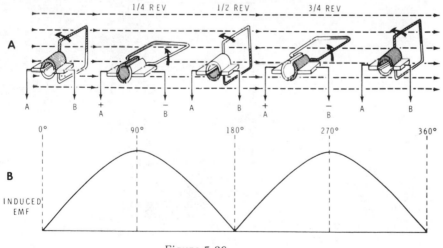

Figure 5-26
Operation of DC generator.

161

MAGNETIC AND
ELECTROMAGNETIC APPLICATIONS

You have already seen two important applications of electromagnetism — the alternator and the DC genterator. It would be difficult to list all of the other applications of magnetic and electromagnetic devices. However, we can examine the operation of some of the more common applications.

Relay

The relay is one of the simplest electromagnetic devices. It is also one of the most useful. Figure 5-27 shows how the relay operates. When the switch is closed, current flows from the battery through the relay coil. The current develops a magnetic field in the core which attracts the armature, pulling it down. This causes the two contacts to close, connecting the generator to the load.

When the switch is opened, the current through the relay coil stops. This allows the magnetic field to collapse. The spring pulls the armature back up opening the contacts and disconnecting the generator from the load.

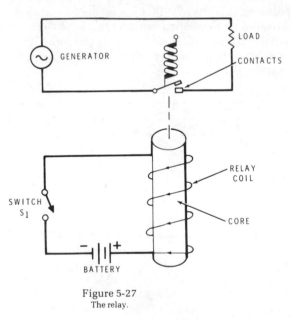

Figure 5-27
The relay.

The relay is used when it is desirable to have one circuit control another. Notice that in Figure 5-27 there are two complete and separate circuits. Because the relay circuit is electrically isolated from the generator circuit, the relay can be used to open and close high-voltage or high-current circuits with relatively little voltage and current in the coil circuit. It is also useful for remote control where the switch is located at one point and the other circuit components are located at a distance. Also, a relay with several contact arms can open and close several circuits at once.

An interesting application of the relay is the door bell shown in Figure 5-28. When the switch is closed, current flows from the negative side of the battery through the switch, the breaker contacts, the two relay coils, and back to the positive side of the battery. The current flow through the relays sets up a magnetic field which attracts the soft-iron armature pulling it down. This pulls the hammer down causing it to strike the bell. The lower breaker contact is attached to the armature. Consequently, when the coils energize, the current path for the coil is broken. Thus, the relay de-energizes and releases the armature. The spring pulls the armature and the hammer upward. This closes the breaker contacts and once again completes the path for current flow through the relays. The operation repeats itself many times each second. Thus the bell is rung as long as the circuit is energized.

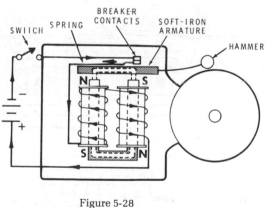

Figure 5-28
The door bell.

163

Reed Switch and Relay

Figure 5-29A shows a magnetic reed switch. It consists of two contacts in a sealed glass container. The contacts are made of a ferromagnetic material and are normally open. However, when a magnet is placed next to the reed switch as shown in Figure 5-29B, the contacts close. The reason for this is that a magnetic field is induced into each contact by the flux lines from the magnet. Thus, each contact becomes a tiny magnet having the polarity shown. At the point where the two contacts are closest, opposite poles exist. These poles are attracted to each other closing the contacts. The reed switch allows us to turn a current on or off by changing the position of a permanent magnet. A practical application of this device will be shown later.

Figure 5-29C shows that the contacts can also be controlled with the field from an electromagnet. When the electromagnet is wound directly on the reed switch, the device is called a reed relay.

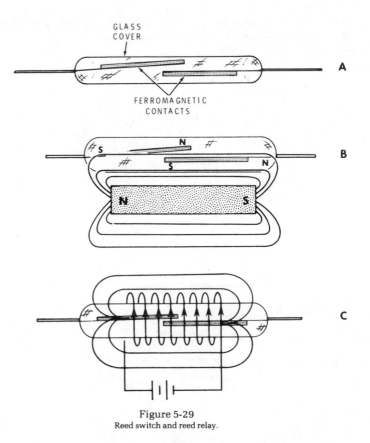

Figure 5-29
Reed switch and reed relay.

164

Record Pickup

Electromagnetic principles are used in many types of recording and playback equipment. The pickup cartridge used in the tone arm of many record players are electromagnetic devices. Figure 5-30 shows the construction of a cartridge called a moving-coil or dynamic pickup. Here a magnetic field is produced by the permanent magnet. A tiny coil is placed in this magnetic field. The core on which the coil is wound is attached to the stylus or needle. The coil is held in place by a flexible grommet. As the needle slips down the spiral groove on the record, it vibrates in response to the variations in the groove. These variations in the groove correspond to the audio tones recorded there. Thus, the needle vibrates at the same rate as the audio tones. Because the coil is connected to the needle, it also vibrates at this rate. The tiny movements of the coil in the magnetic field cause a minute emf to be induced into the coil. The induced emf also varies at the audio rate. This emf can be amplified and used to drive a loudspeaker so that the original audio tone is reproduced.

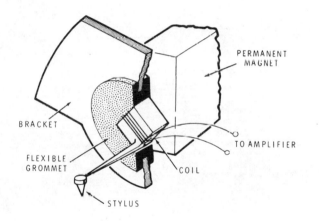

Figure 5-30 Simplified diagrams of a magnetic pickup cartridge.

Loudspeaker

Loudspeakers are used in all types of audio equipment. Most loudspeakers use a moving coil and a permanent magnet. A cutaway diagram of a loudspeaker is shown in Figure 5-31. A permanent magnet establishes a strong stationary magnetic field. A coil which is free to move is placed in this magnetic field. A current which varies at an audio rate is then passed through the coil. The varying current establishes a varying magnetic field around the coil. The varying magnetic field of the coil is alternately attracted and repelled by the stationary field of the permanent magnet. Thus, the coil moves back and forth at the same rate as the varying current. The moving coil is attached to a large cone or diaphragm. As the coil vibrates, the cone also vibrates setting the air around the cone in motion at the same rate. This reproduces the original sound.

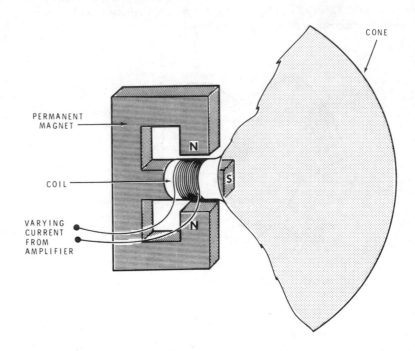

Figure 5-31
The loudspeaker.

166

Magnetic Tape

The tape recorder uses electromagnetic principles to record electronic signals on magnetic tape. The device which actually "writes" the signal on the tape and later "reads" it back is called a record-playback head. It is nothing more than a coil with a ferromagnetic core. Figure 5-32A illustrates the operation in the record mode.

Notice that a tiny air gap exists between the two ends of the core. When current is applied to the coil, a magnetic field is concentrated in this gap. A length of magnetic tape is pulled past the air gap. The plastic tape is covered with a ferromagnetic substance such as iron oxide. The magnetic field surrounding the air gap penetrates the tape magnetizing it at this point. If the current applied to the coil varies at an audio rate, then the magnetic field across the air gap varies at the same rate. Consequently, the magnetic pattern "written" on the tape corresponds to the original audio signal.

To play back the tones recorded on the tape, the process is reversed as shown in Figure 5-32B. The tape is pulled past the air gap so that the core is subjected to the magnetic patterns on the tape. The changing magnetic field induces a tiny emf into the coil windings. When this emf is amplified and applied to a loudspeaker, the original audio tones are reproduced.

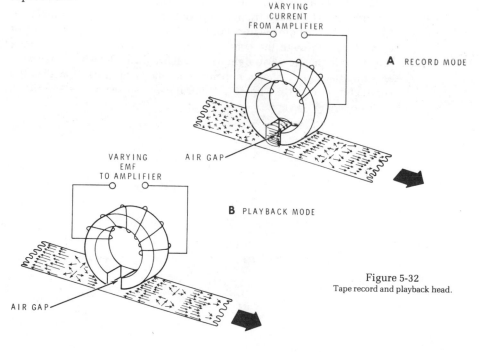

Figure 5-32
Tape record and playback head.

DC Motor

Earlier you saw that a generator converts mechanical energy to electrical energy. A motor does just the opposite; it converts electrical energy to mechanical energy. Figure 5-33A illustrates the principle which makes this possible. Here a current carrying conductor is shown in a magnetic field. This is not an induced current; it flows because the conductor is connected across a battery. Because of the current, a magnetic field is developed around the conductor in the direction shown. This can be verified by the left-hand rule for conductors which was discussed earlier. The magnetic field around the conductor interacts with the field of the permanent magnet. Notice that on one side of the wire, the two magnetic fields have the same direction and they add thus producing a strong magnetic field. On the other side of the conductor the two magnetic fields have opposite directions. Thus, they tend to cancel leaving a weak resultant field at this point. As you can see, the flux lines are much more numerous on one side of the conductor than on the other. Thus, on one side, the lines are bent and are forced very close together. These lines have a natural tendency to straighten and move further apart. However, the only way they can do this is to push the conductor out of the way. Thus, a force is developed which will push the conductor in the direction shown.

There is a rule for determining the direction that the wire will move. It is called the right-hand motor rule, and is illustrated in Figure 5-33B. Using the right-hand (not the left), point the index finger in the direction of the field of the permanent magnet. Point the middle finger in the direction of current flow through the conductor and at a right angle to the index finger. Point the thumb straight up and at a right angle to both the index and middle finger. The thumb now points in the direction that the conductor will move. Applying this rule we can see how a simple dc motor operates.

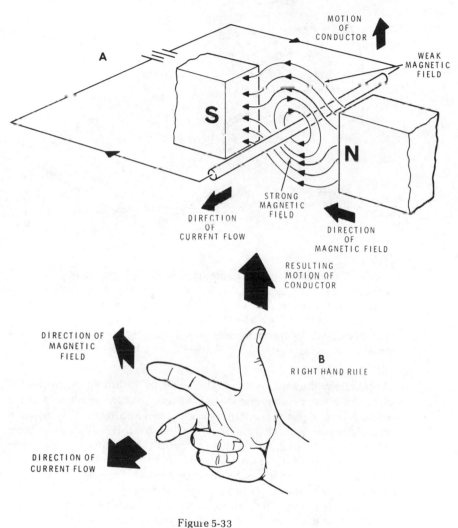

MOTION
OF
CONDUCTOR

A

WEAK
MAGNETIC
FIELD

S

N

DIRECTION
OF
CURRENT FLOW

STRONG
MAGNETIC
FIELD

DIRECTION
OF
MAGNETIC FIELD

RESULTING
MOTION OF
CONDUCTOR

DIRECTION OF
MAGNETIC
FIELD

B
RIGHT HAND RULE

DIRECTION OF
CURRENT FLOW

Figure 5-33
Rules for motor action.

169

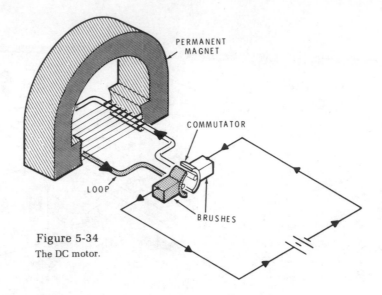

PERMANENT
MAGNET

COMMUTATOR

LOOP

BRUSHES

Figure 5-34
The DC motor.

A simplified diagram of the DC motor is shown in Figure 5-34. Notice the similarity to the DC generator described earlier. However, there are two important differences. With the generator, the loop was turned by an outside mechanical force. Here the loop turns because of the motor action just discussed. In the generator, a DC voltage was produced at the brushes. Here, an external DC voltage from a battery is applied to the brushes.

Current flows through the loop as indicated by the arrows. Applying the right-hand motor rule to the side of the loop near the south pole of the magnet, we find that the conductor tends to move up. If we apply the same rule to the side near the north pole, we find that this side tends to move down. Thus, the loop rotates in a counterclockwise direction. After one half cycle of revolution, the two sides reverse positions. Nevertheless, current still flows in the same direction through the side closest to the south pole. Whichever side of the loop appears at this point, the resulting motion is always up. The upward force at the south pole and the downward force at the north pole causes the loop to constantly rotate.

The simple motor shown here is not practical because a single loop of wire is used as an armature. Real motors use hundreds of turns of wire so that a very strong torque is developed.

170

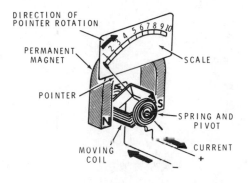

DIRECTION OF
POINTER ROTATION

PERMANENT
MAGNET

POINTER

SCALE

SPRING AND
PIVOT

MOVING
COIL

CURRENT
+

Figure 5-35
The moving-coil meter movement.

Meter

The same motor action described above is used in the moving-coil meter movement. Figure 5-35 shows a simplified diagram of this device. Like the motor it has an armature which is free to revolve in the field of a permanent magnet. However, in the meter movement, the motion is restricted by one or more springs. When current flows through the coil, it establishes a magnetic field the strength of which is directly proportional to the current. The motor action causes the coil to rotate. However, the restraining springs prevent the coil from rotating more than about 90°. A pointer is attached to the coil. As the coil rotates, it moves the pointer in front of a scale. The more current that flows, the further the coil will rotate and the further up the scale the pointer will move. The scale reading is directly proportional to the amount of current that flows through the coil. Therefore, the scale can be marked off in amperes, milliamperes, or even microamperes. This type of meter movement is used in most ammeters, voltmeters, and ohmmeters.

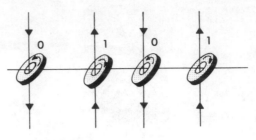

Figure 5-36   Four cores of a
computer memory.

Computer Memories

Computers use a variety of electromagnetic devices to store information.
From the start, one of the most used storage techniques has been magnetic
cores. These are tiny little doughnut-shaped pieces of ferrite material
which can be magnetized in either of two directions. As shown in Figure
5-36, wires are strung through the holes in the cores. By applying current
to these wires in the appropriate direction, the cores can be magnetized in a
certain pattern. A clockwise magnetic field can arbitrarily be called 1 while
a counterclockwise field can be called 0. The cores in Figure 5-36 have the
pattern 0101. Patterns of 1's and 0's can be used to represent numbers,
letters of the alphabet, and punctuation marks. For example, one popular
computer code uses seven digit patterns. The letter A is represented by
1000001; the number 6 by 0110110; and the question mark (?) by
01111111. Thus, if we are willing to use enough cores, the entire contents
of this course could be stored in a core memory using seven digit patterns
of 1's and 0's.

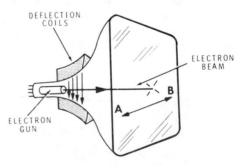

Figure 5-37   Magnetic deflection
of the electron beam in a TV picture
tube.

172

Magnetic Deflection of Electron Beams

We have seen that a current carrying conductor is deflected (moved) by a magnetic field. However, it is not the conductor which is deflected but the electrons traveling through the conductor. Since the electrons are confined to the wire, the conductor moves also.

In some cases streams of electrons are not confined to a wire but travel through empty space. Such a beam can be deflected in the same way that a current carrying conductor is deflected. There are many practical applications of this principle. The most familiar is the TV picture tube like the one shown in Figure 5-37. A device called an electron gun produces a narrow beam of electrons which is fired at the TV screen. Wherever the beam hits the phospher screen, light is given off. By moving the beam over the entire surface of the screen while varying its intensity, a picture can be drawn.

Two magnetic fields are used to deflect the beam. One moves the beam back and forth across the screen over 15,000 times each second. Another moves the beam up and down the screen 30 times each second. The result is that 30 complete pictures consisting of about 500 lines each are drawn each second.

The horizontal deflection coils are shown in Figure 5-37. When current flows in one direction through these coils, a magnetic field having the direction shown is produced. Using the right-hand motor rule, we can see that this field will deflect the beam toward point A. If the direction of the field is reversed, the beam will deflect toward point B. To deflect the beam in a vertical direction, vertical deflection coils are placed on the sides of the picture tube. This principle is used in radar sets and TV cameras as well as TV receivers.

SUMMARY

Action-at-a-distance is explained in terms of a field. Because a magnet can affect objects at a distance, it is said to be surrounded by a magnetic field. The field is assumed to be made-up of lines of force called flux lines.

Magnets have north and south poles. The flux lines leave the magnet at the north pole and enter the magnet at the south pole forming complete loops. The flux lines cannot cross each other and tend to form the smallest possible loops.

Like magnetic poles repel; unlike poles attract.

All materials are classified as either ferromagnetic, paramagnetic, or diamagnetic. Ferromagnetic materials are strongly attracted by magnetic fields; paramagnetic materials are only slightly attracted; and diamagnetic materials are slightly repelled. Generally, paramagnetic and diamagnetic materials are considered to be non-magnetic.

When an electrostatic charge moves, a magnetic field is developed. The electron spins on its axis producing a magnetic field. In most atoms, electrons with opposite spins pair off so that their magnetic fields cancel. However, in iron, nickel and cobalt, the two valence electrons spin in the same direction and thus their fields add.

These magnetic atoms bunch in tiny groups called domains. Normally, these domains are arranged haphazardly and their fields cancel. However, when subjected to a magnetic field, the domains align in the same direction creating a magnet.

Magnetism and electricity are closely related. Current flow produces a magnetic field and a moving magnetic field can produce current flow.

The direction of the magnetic field caused by current flow can be determined by the left-hand rule for conductors. This rule states: Grasp the conductor in your left hand with the thumb pointing in the direction of current flow. Your fingers now point in the direction of the flux lines.

The magnetic field around the conductor can be strengthened and concentrated by winding the conductor as a coil. The result is called an electromagnet which has many of the characteristics of a permanent magnet.

The north pole of the electromagnet can be determined by the left-hand rule for coils. It states: Grasp the coil in your left hand with your fingers wrapped around it in the direction that current is flowing. Your thumb now points toward the north pole of the coil.

174

Several magnetic quantities are important. *Permeability* is the ease with which a substance accepts lines of force. Its reciprocal is called *reluctance*. *Flux* is the total lines of force around a magnet. *Flux density* refers to the amount of flux per unit of area. *Magnetomotive force* is the force which produces the flux in a coil. *Field intensity* is the amount of mmf per unit of length of the coil.

When one body has an electrostatic or magnetic field, it can induce a change in another body without actually touching the other body. This is called induction.

A magnet can induce a magnetic field into a ferromagnetic body without touching it. This is called magnetic induction. When the magnet is taken away, a magnetic field will remain in the ferromagnetic body. This is called *residual magnetism*. The ability of a substance to retain a magnetic field after the magnetizing force has been removed is called *retentivity*.

When a conductor moves across a magnetic field, an emf is induced in the conductor. This is called *electromagnetic induction*.

The magnitude of the induced emf is proportional to the rate at which the conductor cuts the magnetic lines of force. The more lines per second that are cut, the higher the induced emf will be.

The polarity of the induced emf can be determined by the left-hand rule for generators. Using this rule, the thumb is pointed in the direction that the conductor is moving. The index finger is pointed in the direction of the flux lines. If the middle finger is now placed at right angles to the thumb and index finger, it points in the direction in which current will flow through the conductor.

A device which uses electromagnetic induction to convert mechanical energy to electrical energy is called a generator. A generator may produce AC or DC depending on how it is constructed. An AC generator is called an alternator. The DC generator uses commutators and brushes to convert AC to DC.

There are many other devices which use magnetic or electromagnetic principles.

The *relay* uses an electromagnet to close switch contacts.

The *record pickup* uses a magnet and moving coil to convert stylus vibrations into voltage variations which can be amplified. The *loudspeaker* uses a similar arrangement to convert these amplified voltage variations back to sound.

The tape recorder uses an electromagnet as a *record* head. Current variations are applied to the electromagnet which creates corresponding magnetic patterns on a length of tape. In the record mode, the head converts to a tiny generator. The magnetic variations on the tape induce tiny voltage variations into the coil. These voltage variations are amplified and applied to the loudspeaker where they are converted back to sound.

The *motor* converts electrical energy to mechanical energy. Current is passed through a coil producing an electromagnetic field. This field interacts with a permanent magnet's field. As a result, the coil is forced to move. By proper design, a constant circular motion is achieved.

The meter works on the same principle as the motor. Here the movement of the coil is restricted by springs. A pointer attached to the coil moves in front of a scale indicating the current through the meter.

Motor action is also used to deflect the electron beam in TV receivers, TV cameras, and radar indicators.

The direction of movement of the coil in the motor or meter, or the direction of deflection of the electron beam can be determined by the right-hand motor rule. Using this rule, the thumb, index finger, and middle finger of the right hand are held at right angles to each other. If the index finger is pointed in the direction of the magnetic field and the middle finger is pointed in the direction of electron flow, the thumb will point in the direction which the conductor (or electron beam) will move.

Unit 6

ELECTRICAL MEASUREMENTS

INTRODUCTION

This unit deals with electrical measurements. It explains the construction and operation of the most commonly used electrical measuring instrument – the volt-ohm-milliammeter (VOM). As the name implies, this instrument is used to measure voltage, resistance, and current. The heart of this instrument is a moving-coil meter movement like the one discussed in the previous unit on magnetism. In this unit we will take a closer look at this type of meter movement and see how it can be used to measure current, voltage, and resistance.

THE METER MOVEMENT

The heart of the volt-ohm-milliammeter (VOM) is the meter movement. The most popular type of meter movement is the **permanent-magnet**, moving-coil movement discussed in the previous unit. This device is also called the d'Arsonval movement after its inventor Arsene d'Arsonval. The first version, introduced in 1882 was called a galvanometer. It was quite delicate and somewhat crude. In 1888, Edward Weston introduced an improved version of the device which is similar to the designs used today.

Construction

Figure 6-1 shows the construction of the meter movement. Several important parts are listed. Let's discuss these parts in detail starting with the permanent magnet.

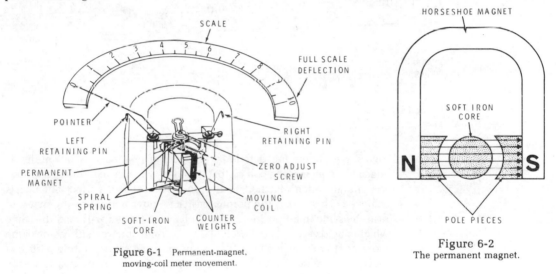

Figure 6-1   Permanent-magnet,
moving-coil meter movement.

Figure 6-2
The permanent magnet.

Figure 6-2 shows the permanent magnet system. A horse shoe magnet produces the stationary magnetic field. To concentrate the magnetic field in the area of the moving coil, pole pieces are added to the magnet. These are made of soft iron and have a very low reluctance. Consequently, the lines of flux tend to concentrate in this area as shown. Also, a stationary soft-iron core is placed between the pole pieces. Enough space is left between the pole pieces and the core so that the moving coil can rotate freely in this space. As you can see, the pole pieces and core restrict most of the flux to the area of the moving coil.

Figure 6-3 shows how the moving coil fits around the soft-iron core. The coil consists of many turns of extremely fine wire on an aluminum frame. The aluminum frame is very light so that little torque is needed to move it. The two ends of the coil connect to the leads of the ammeter, voltmeter, or ohmmeter.

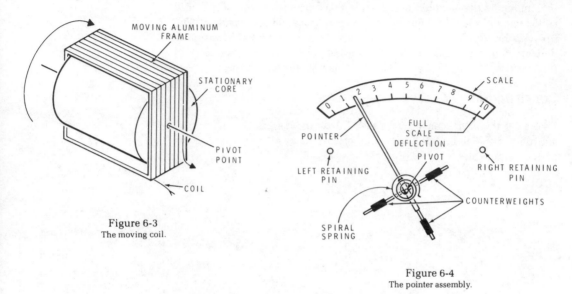

Figure 6-3
The moving coil.

Figure 6-4
The pointer assembly.

Figure 6-4 shows the details of the pointer assembly. The pointer is attached to the moving coil so that it moves when the coil does. Counterweights are often attached to the pointer so that a perfect balance is achieved. This makes the pointer easier to move and helps the meter to read the same in all positions. A well balanced meter will read the same whether held vertically or horizontally. Retaining pins on either side of the movement limit the distance that the pointer and other rotating parts can move. Two spiral springs at opposite ends of the moving-coil force the pointer back to the zero position when no current is flowing through the coil. In most movements, the spiral springs are also used to apply current to the moving coil. The two ends of the coil connect to the inner ends of the spiral spring. The outer end of the rear spring is fixed in place. However, the outer end of the front spring connects to a zero adjust screw. This allows you to set the pointer to exactly the zero point on the scale when no current is flowing through the coil.

The moving coil, pointer, and counter-weight rotate around a pivot point. To hold the friction to an absolute minimum, jeweled bearings are used at this point just as they are in a fine watch.

180

Operation

Now that you have an idea of how the meter is constructed, let's see how it operates. In the previous unit you learned that a conductor is deflected at a right angle to a stationary magnetic field if current flows through the conductor. This is the principle of the DC motor. You also learned a rule, called the right-hand motor rule, which describes this action. Figure 6-5 illustrates this rule and the motor action which causes the meter to deflect. An end view of one turn of the moving-coil is shown. Current is forced to flow through the coil so that current flows "out of the page" on the left. Applying the right hand rule to the coil at this point, we find that the coil is forced up on the left and down on the right. This forces the pointer to move up scale or in the clockwise direction.

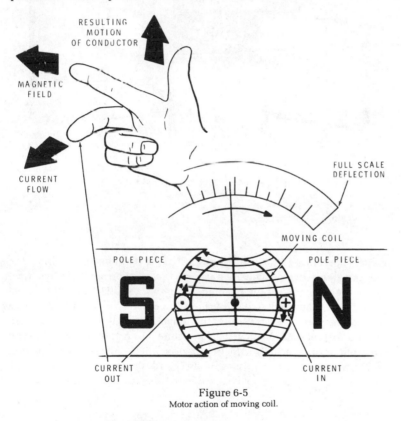

Figure 6-5
Motor action of moving coil.

The amount of torque produced by this tiny "motor" is proportional to the magnitude of the current which flows through the moving coil. The more current, the greater the torque will be and the further the pointer will be deflected.

181

Meter movements are rated by the amount of current required to produce full-scale deflection. For example, a 50 microampere meter movement deflects full scale when only 50 microamperes of current flows through it. The 50 μA meter movement is one of the most commonly used types of d'Arsonval movements. The 100μA and 200μA movements are also popular.

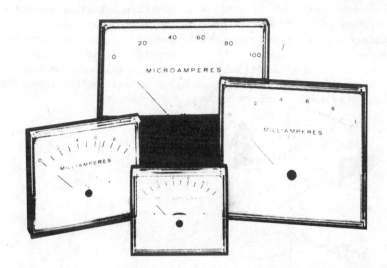

Meter movements come in different shapes, sizes, and sensitivities. Courtesy Weston.

We can use Figure 6-5 to illustrate an important characteristic of the d'Arsonval meter movement. We assume that the current is always flowing in the same direction through the moving coil. That is, a direct current (DC) is applied to the coil. This movement will work fine as long as the current is direct. However, the movement will not respond properly to an alternating current (AC). Each time the current reverses, the coil will attempt to reverse its direction of deflection. If the current changes direction more than a few times each second, the coil cannot follow the changes. Thus, AC must not be applied to this type of meter movement.

Taut-Band Movement

An important variation of the d'Arsonval movement is the taut-band meter movement. Figure 6-6 is a simplified diagram which shows the construction of this type of movement. The moving coil is suspended by two tiny stretched metal bands. One end of each band is connected to the moving coil while the other end is connected to a tension spring. The purpose of the springs is to keep the bands pulled tight. The bands replace the pivots, bearings, and spiral springs used in the conventional d'Arsonval movement. This not only simplifies the construction of the meter, it also reduces the friction to practically zero. Consequently, the taut-band movement can be made somewhat more sensitive than the movement discussed earlier. Taut-band instruments with 10 μA movements are available.

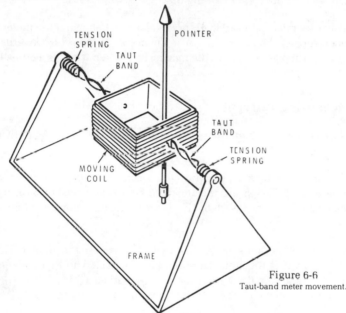

Figure 6-6
Taut-band meter movement.

The taut-bands serve several purposes. First, they suspend the coil in such a way that the friction is nearly zero. When current is applied, the coil rotates and the bands are twisted. When current is removed, the bands untwist returning the pointer to the zero position. The bands also serve as the current path to and from the coil.

The taut-band movement has several advantages over the original d'Arsonval type. As we have seen, it is generally more sensitive. It is also more rugged and durable. Mechanical shocks simply deflect the tension springs which can then bounce back to their original positions. The instrument remains more accurate for the same reason. Because of these advantages, the taut-band movement is becoming increasingly popular.

183

THE AMMETER

The meter movements discussed in the previous section are basically current meters. That is, they deflect when current flows through them. In each case, the moving coil consists of many turns of extremely fine wire. Current is carried to the coil via the fragile spiral springs or the taut-bands of the movement. Because of the delicate nature of the coil and the springs or bands, care must be taken not to feed excessive current through the movement. The current necessary for full-scale deflection will not harm the movement, but a 100% overload might. The coil may burn out; the spring may be damaged; or the aluminum needle may be bent if driven too hard against the right retaining pin.

Also, care must be taken to observe polarity when using the meter movement. A reverse current will cause the needle to deflect backwards. If the current is too great, the needle may be bent when it bangs against the left retaining pin.

Measuring Current

The rules for using the ammeter were discussed earlier. Let's briefly review these rules.

First, the ammeter must be connected in series with the current which is to be measured. This means that the circuit under test must be broken so that the ammeter can be inserted. This is the prime disadvantage of the ammeter.

Second, polarity must be observed when connecting the meter. This simply means that the ammeter should be connected so that it deflects up scale. The terminals of most meters are marked with − and +. Simply connect the meter so that current flows into the − terminal.

Third, the current ratings of the meter must not be exceeded.

Increasing the Range of the Ammeter

Each meter movement has a certain current rating. This is the current which will cause full-scale deflection. For example, an inexpensive meter movement may have a current rating of 1 milliampere. To obtain a usable reading, the current through the movement cannot be more than 1mA. By itself, the movement has a single usable range of 0 to 1 mA.

Obviously, the meter would be much more useful if it could measure currents greater than 1 mA as well as those less than 1 mA. Fortunately, there is an easy way to convert a sensitive meter movement to a less sensitive current meter. We do this by connecting a small value resistor in parallel with the meter movement. The resistor is called a *shunt*. Its purpose is to act as a low resistance path around the movement so that most of the current will flow through the shunt and only a small current will flow through the movement.

Figure 6-7A illustrates a 1 mA meter movement connected across a low resistance shunt to form a higher range ammeter. The range will depend on how much current flows through the shunt. In Figure 6-7B, the current applied to the ammeter is 10mA. However, only 1mA of this flows through the meter movement. The other 9mA flows through the shunt. Thus, to convert the 1mA movement to a 0-10mA meter, the shunt must be chosen so that 9/10 of the applied current flows through the shunt. Once this is done, the full scale position on the scale indicates 10mA since this is the amount of current that must be applied before full scale deflection is reached.

If the value of the shunt is made smaller, the meter can indicate even higher values of current. Figure 6-7C shows the requirements necessary to measure 100 mA. Here 99mA or 99 percent of the applied current must go through the shunt. Thus, the resistance of the shunt must be much smaller than the resistance of the meter movement.

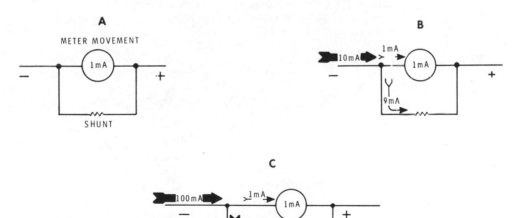

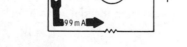

Figure 6-7   Increasing the current range of the ammeter.

Computing the Shunt Resistance

To determine the proper value of the shunt resistor, we must first know something of the characteristics of the meter movement. In the earlier example, we know that full scale deflection requires 1mA. However, we must also know either the resistance of the meter movement or the voltage dropped by the movement when the current is 1mA. Of course, if we know one, we can compute the other.

The resistance value of the meter movement is given in the manufacturer's literature, catalog, or operating instructions. Often it is printed right on the meter movement itself. Let's assume that the 0-1mA movement has a resistance of 1000 ohms or 1 kilohm. In this case, 1mA of current causes a voltage drop across the meter movement of:

$$E = IR$$
$$E = 1mA \times 1\ k\Omega$$
$$E = .001A \times 1000\ \Omega$$
$$E = 1\ volt$$

If we refer back to Figure 6-7, we see that this is the voltage developed across the meter movement in each of the examples shown. Since the shunt resistance is connected in parallel with the meter movement, this same voltage must be developed across the shunt. This means that in the example shown in Figure 6-7B, the 9mA current must develop 1 V across the shunt. Using Ohm's law, we can now compute the value of the shunt since we now know the current and voltage. Thus, the value of the shunt should be:

$$R = \frac{E}{I}$$

$$R = \frac{1V}{9mA}$$

$$R = \frac{1V}{.009A}$$

$$R = 111\ ohms$$

This is the resistance necessary to shunt 9mA around the meter when a current of 10mA is flowing in the circuit. However, the shunt works equally well when the movement is indicating half-scale or 0.5mA. Again, the voltage across the meter can be calculated by Ohm's law:

$$E = IR$$
$$E = 0.5mA \times 1K\Omega$$
$$E = 0.0005A \times 1000\Omega$$
$$E = 0.5\ volt$$

Since the voltage across the meter is the same as that across the 111 ohm resistor, the current through the shunt is:

$$I = \frac{E}{R}$$

$$I = \frac{0.5V}{111 \ \Omega}$$

$$I = .0045A \text{ or } 4.5mA$$

As you can see, nine-tenths of the current still flows through the shunt while only one-tenth flows through the movement. Thus, the movement indicates 0.5mA when 5mA of current flows in the circuit. The meter scale is marked off 0 through 10 rather than 0 through 1 and a 111 ohm resistor is connected across the meter movement. This converts the circuit to a 0 – 10mA current meter.

Interestingly enough, we can find the value of shunt required in another way. Refer to Figure 6-7B again. We know that 9 mA must flow through the shunt so that 1mA is left to flow through the meter movement. In order for the shunt to conduct 9 times as much current as the meter, its resistance must be only 1/9 that of the meter resistance. Now since the meter resistance is 1000 ohms, the shunt resistance must be $\frac{1000\Omega}{9}$ or 111 ohms.

Let's try applying these two methods to the situation shown in Figure 6-7C. Once again the current through the meter movement is 1mA. Thus, the voltage drop across the movement and across the shunt is still 1 volt. This allows us to compute the value of the shunt:

$$R = \frac{E}{I}$$

$$R = \frac{1 \ V}{99 \ mA}$$

$$R = \frac{1 \ V}{.099A}$$

$$R = 10.1 \text{ ohms}$$

We arrive at this same answer by reasoning that the shunt resistance must be 1/99 that of the meter resistance since the shunt conducts 99 times as much current. Thus the shunt resistance must be:

$$\frac{1000\Omega}{99} = 10.1 \text{ ohms.}$$

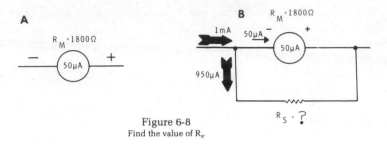

Figure 6-8
Find the value of R_s.

Let's try another example. Figure 6-8A shows a 0-50μA meter movement with a resistance of 1800 ohms. What value shunt is required to construct a 0-1mA meter? Figure 6-8B shows the current distribution. Notice that 1mA or 1000μA of current flows through the circuit. However, only 50μA can flow through the meter movement. The remaining 950μA must flow through the shunt. Since we know the resistance of the meter movement (1800Ω) and the current through it (50μA), we can compute the voltage across it:

$$E = IR$$
$$E = 50 \ \mu A \times 1800 \ \Omega$$
$$E = 0.00005A \times 1800 \ \Omega$$
$$E = 0.09 \text{ volts}$$

Because the shunt is in parallel with the meter movement, this same voltage is developed across the shunt. Thus, we can compute the shunt value:

$$R_S = \frac{E}{I}$$
$$R_S = \frac{0.09V}{950\mu A}$$
$$R_S = \frac{0.09V}{0.00095A}$$
$$R_S = 94.7\Omega$$

As before, we can arrive at this same answer by reasoning that 95% of the current flows through the shunt. This means that 95/5 or 19 times as much current flows through the shunt as through the meter movement. Because the shunt conducts 19 times as much current, the resistance of the shunt must be 1/19 that of the meter movement. Thus the shunt resistance must be:

$$\frac{1800\Omega}{19} = 94.7\Omega.$$

188

Ammeter Accuracy

Every meter movement has a certain accuracy associated with it. The accuracy is specified as a *percentage of error at full scale deflection*. Accuracies of ±2% or ±3% of full-scale are common for good quality instruments. Figure 6-9 illustrates what is meant by ±3% of full scale. The scale shown is a 100mA current scale. Remember, the meter accuracy refers to full scale deflection. At full scale, ±3% equals ±3mA. For this meter, a current of exactly 100mA could cause the meter to read anywhere from 97mA to 103mA. Another way to look at it is that a meter reading of exactly 100mA might be caused by an actual current of from 97mA to 103mA.

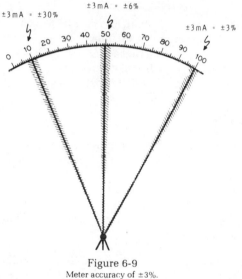

Figure 6-9
Meter accuracy of ±3%.

As you can see a ±3% accuracy means that the reading may be off by as much as ±3 mA at full scale. More importantly, it means that the reading may be off by as much as ±3mA at any point on the scale. For example, when the meter indicates 50mA, the actual current may be anywhere from 47mA to 53mA. Thus, at half-scale the accuracy is no longer ±3%; it is now ±6%. By the same token for an indicated current of 10mA, the actual current may be anywhere from 7mA to 13mA. Here, the accuracy is only ±30%.

Because meter accuracy is specified in this manner, the accuracy gets progressively worse, as we move down the scale. For this reason, current measurements will be most accurate when a current range is selected that will cause near full scale deflection of the meter. The nearer full scale, the more accurate the reading will be.

THE VOLTMETER

The basic meter movement can be used to measure voltage as well as current. In fact every meter movement has a certain voltage rating as well as current rating. This is the voltage which will cause full scale deflection. Of course, the voltage rating is determined by the current rating and the meter resistance. For example, a 50μA meter movement which has a resistance of 2000Ω deflects full scale when connected across a voltage of:

$$E = IR$$
$$E = 50\ \mu A \times 2000\ \Omega$$
$$E = 0.1 \text{ volt}$$

That is, the meter movement alone could be used to measure voltages up to 0.1 volt. Thus, the meter scale can be calibrated from 0 to 0.1 volt. However, if the meter movement is connected across a much higher voltage such as 10 volts, it may be damaged. Obviously, to be practical, we must extend the voltage range of the basic meter movement.

Extending the Range

We have seen that a 50μA, 2000 ohm meter movement can withstand a voltage of 0.1 volt without exceeding full scale. To extend the range, we must insure that the voltage across the meter does not exceed 0.1 volt when the meter movement is connected across a higher voltage. We do this by connecting a resistor in *series* with the meter movement as shown in Figure 6-10. The resistor is called a *multiplier* because it multiplies the range of the meter movement.

The purpose of the multiplier resistor is to limit the current which flows through the meter movement. For example, in the voltmeter shown in Figure 6-10, the current through the meter movement must be limited to 50μA. Another way to look at it is that the multiplier must drop all the voltage applied to the voltmeter except the 0.1 volt allowed across the meter movement. For example, if the range is to be extended to 10 volts, then the multiplier must drop 10V − 0.1V = 9.9 volts.

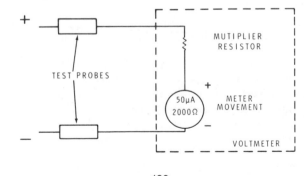

Figure 6-10 A voltmeter is formed by connecting a multiplier resistor in series with a microammeter.

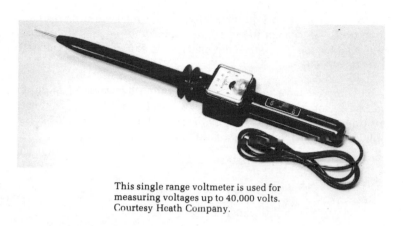

This single range voltmeter is used for
measuring voltages up to 40,000 volts.
Courtesy Heath Company.

Calculating the Multiplier

We have seen that the value of the multiplier must be high enough to limit
the current to the full-scale current rating of the meter movement for any
applied voltage. If we keep this in mind, we can easily calculate the
required value of the shunt for any voltage range.

Let's assume we wish to convert the $50\mu A$, 2000 ohm meter movement to a
10-volt voltmeter by adding a multiplier in series. Obviously, a current of
only $50\mu A$ must flow when the voltmeter is connected across 10 volts.
Thus, the total resistance of the voltmeter must be:

$$R_{total} = \frac{E_{\text{full-scale}}}{I_{\text{full-scale}}}$$

$$R_{total} = \frac{10V}{50\mu A}$$

$$R_{total} = 200,000\Omega$$

However, the meter movement itself has a resistance of 2000 ohms. Thus,
the multiplier must have a value of 200,000 Ω − 2000 Ω = 198,000 Ω or
198 KΩ.

This means that the basic $50\mu A$, 2000Ω meter movement can now measure
0 to 10 volts because 10 volts must be applied to cause full-scale deflection.
From a voltage standpoint, the multiplier resistor drops 99% of the applied
voltage. That is, for an applied voltage of 10 volts, the multiplier drops:

$$E = IR$$
$$E = 50\ \mu A \times 198,000\ \Omega$$
$$E = 9.9\ volts$$

This leaves 0.1 volt across the meter. Because, the total voltmeter resistance is 100 times larger than the meter resistance, the range of the meter is multiplied by 100. Of course, the meter scale should now be calibrated from 0 to 10 volts.

To be sure you have the idea, let's determine the value of multiplier required to convert the same meter movement to a 0-100 volt voltmeter. This time, the multiplier must limit the current to $50\mu A$ when 100 volts is applied. Thus, the total resistance of the voltmeter must be:

$$R_{total} = \frac{E_{full\text{-}scale}}{I_{full\text{-}scale}}$$

$$R_{total} = \frac{100V}{50\mu A}$$

$$R_{total} = 2,000,000 \text{ ohms or } 2 \text{ M}\Omega$$

Here again, 2000 ohms are supplied by the meter movement. Thus, the value of the multiplier must be:

$$R_{multiplier} = R_{total} - R_{meter}$$

$$R_{multiplier} = 2,000,000 \text{ } \Omega - 2000 \text{ } \Omega$$

$$R_{multiplier} = 1,998,000 \text{ } \Omega \text{ or } 1.998 \text{ M}\Omega$$

Notice that 1.998MΩ is extremely close to 2MΩ. The difference is so slight that we probably would not notice any difference in deflection regardless of which value was used. In this case, a standard value 2MΩ resistor would probably be used instead of the 1.998 MΩ value which is not easily obtained.

Multiple-Range Voltmeters

A practical voltmeter has several ranges. One arrangement for achieving multiple ranges is shown in Figure 6-11. Here, the voltmeter has four ranges which may be selected by the range switch. Again, the $50\mu A$, 2000Ω meter-movement is used. On the 0.1-volt range no multiplier is required since this is the voltage rating of the meter movement itself.

On the 1-volt range, R_1 is switched in series with the meter movement. The value of R_1 is given as 18KΩ. Using the procedure outlined earlier, verify that this is the proper value multiplier required.

Notice that on the 10-volt and 100-volt ranges, the multiplier values computed earlier are switched in series with the meter movement.

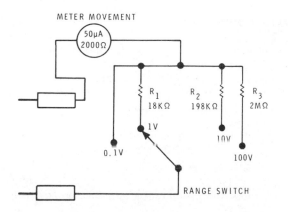

Figure 6-11
Multirange voltmeter

RANGE SWITCH

Figure 6-12 shows another arrangement sometimes used when several ranges are required. On the 0.1-volt range, no multiplier is required. On the 1-volt range an 18 KΩ multiplier is switched in series with the meter movement. Up to this point the arrangement is similar to that shown in Figure 6-11. However, here the similarity ends. On the 10-Volt range, R_2 is switched in series with R_1. Thus, the total resistance in series with the meter movement is 18 KΩ + 180 KΩ = 198 KΩ. Notice that this is the same value of multiplier used on the 10-volt range in Figure 6-11. The only difference is that in Figure 6-11 a single 198 KΩ resistor is used while in Figure 6-12 two resistors having a total resistance of 198 KΩ are used.

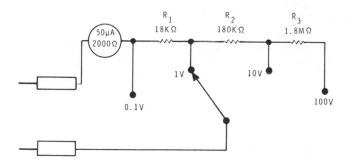

Figure 6-12   Here the multiplier resistors are connected in series.

On the 100-volt range, R_3 is switched in series with R_1 and R_2. Thus, the total multiplier resistance is 18KΩ + 180KΩ + 1.8MΩ = 1.998 MΩ. You will recall that this is the exact multiplier value that we computed earlier for the 100 volt range.

Sensitivity (Ohms per Volt)

An important characteristic of a voltmeter is its *sensitivity*. Sensitivity can be thought of as the amount of current required to produce full scale deflection of the meter movement. For example, a 50μA meter movement is more sensitive than a 1mA meter movement because less current is required to produce full scale deflection.

However, sensitivity is more often defined in another way. Sensitivity is normally expressed in ohms per volt (also written ohms/volt). The more sensitive the meter is, the higher the ohms-per-volt rating will be. The sensitivity in ohms per volt of any voltmeter can be determined simply by dividing the full-scale current rating of the meter movement into 1 volt. That is:

$$\text{Sensitivity} = \frac{1 \text{ volt}}{I_{full-scale}}$$

Thus, the sensitivity of a voltmeter which uses a 50μA meter movement is:

$$\text{Sensitivity} = \frac{1 \text{ volt}}{I_{full-scale}}$$

$$\text{Sensitivity} = \frac{1 \text{ volt}}{50 \, \mu\text{A}}$$

$$\text{Sensitivity} = 20,000 \text{ ohms per volt}$$

This means that on the 1 volt range, the voltmeter has a total resistance of 20,000 ohms. You can prove this by referring back to Figures 6-11 or 6-12. On the one volt range the multiplier has a value of 18 KΩ and the meter movement has a resistance of 2 KΩ. Consequently, the total resistance is 20,000 ohms.

The sensitivity is determined solely by the full-scale current rating of the meter movement. Thus, it has the same sensitivity regardless of the range used. Consequently, on any range, the voltmeters shown in Figure 6-11 and 6-12 have a resistance of 20,000 ohms $\times$ V where V is the selected full-scale voltage range. Thus, on the 10V range, the total voltmeter resistance is 20,000 ohms $\times$ 10 = 200,000 ohms.

Let's try another example. What is the sensitivity of a voltmeter which uses a 1mA meter movement? Remember, sensitivity is determined by dividing the full-scale current into 1 volt. That is:

$$\text{Sensitivity} = \frac{1 \text{ volt}}{I_{full-scale}}$$

$$\text{Sensitivity} = \frac{1V}{1 \text{ mA}}$$

$$\text{Sensitivity} = 1000 \text{ ohms per volt}$$

What would the total resistance of this voltmeter be on the 5 volt range?
The resistance would be 1000 ohms × 5 = 5000 ohms.

Loading Effect of Voltmeters

An unfortunate aspect of electronics is that the act of measuring an electrical quantity often changes the quantity that we are attempting to measure. When measuring voltage, we must connect a voltmeter across the circuit under test. Since some current must flow through the voltmeter, the circuit behavior is modified somewhat. Often the effects of the voltmeter can be ignored, especially if the meter has a high ohms/volt rating. However, if the voltmeter has a low ohms/volt rating or the circuit under test has a high resistance, the effects of the meter cannot be ignored.

Consider the circuit shown in Figure 6-13. Here two 10 KΩ resistors are connected in series across a 6-volt battery. Since the resistors are the same value, each drops one half of the applied voltage or 3 volts. Thus, we would expect a voltmeter to read 3 volts if it were connected across either resistor. However, if the voltmeter has a low ohms/volt rating, the actual reading may be very inaccurate. Figure 6-14A shows the same circuit with a low-sensitivity voltmeter connected across R_2. The voltmeter has a sensitivity of 1000 ohms/volt. Since we expect the voltage across R_2 to be about 3 volts, the voltmeter is on the 0-10 volt range. Thus, its resistance (R_m) is 1000 Ω × 10 = 10,000 Ω. Because R_m is in parallel with R_2, the total resistance is reduced. The parallel resistance (R_A) of the meter and R_2can be computed as we learned in an earlier assignment:

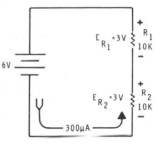

Figure 6-13
R_2 drops 3 volts.

$$R_A = \frac{R_2 \times R_m}{R_2 + R_m}$$

$$R_A = \frac{10,000 \times 10,000}{10,000 + 10,000}$$

$$R_A = \frac{100,000,000}{20,000}$$

$$R_A = 5000\Omega \text{ or } 5 \text{ K}\Omega$$

Therefore, the circuit shown in Figure 6-14A reduces to the circuit shown in Figure 6-14B. Notice how this upsets the operation of the circuit. The total series resistance of R_1 and R_A is now only 15KΩ instead of 20KΩ. This allows more current to flow:

$$I = \frac{E}{R} = \frac{6V}{15,000\Omega} = 0.000400A \text{ or } 400\mu A.$$

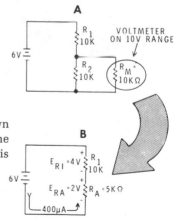

Figure 6-14   The 1000 ohms-per-volt meter loads down the circuit causing an inaccurate reading.

195

Thus, the current increases from its previous value of 300μA to a new value of 400μA. The voltage distribution also changes since R_1 is now larger than R_A. The voltage dropped by R_A is:

$$E_{RA} = I \times R_A$$
$$E_{RA} = 0.0004A \times 5000 \ \Omega$$
$$E_{RA} = 2V$$

The voltage dropped by R_1 increases to:

$$E_{R1} = I \times R_1$$
$$E_{R1} = 0.0004A \times 10,000 \ \Omega$$
$$E_{R1} = 4V$$

Thus, instead of reading 3 volts as we would expect, the meter measures only 2 volts across R_2. This is an inaccuracy of 33%. This effect is called *loading*. We say that the meter is loading down the circuit causing the voltage across R_2 to decrease. The loading effect becomes noticeable when the resistance of the meter approaches that of the resistor across which the meter is connected. For example, if the resistance of the meter were made 10 times that of R_2, then the loading effect would be barely noticeable.

Figure 6-15 shows the same circuit with a 20,000 ohms/volt meter connected across R_2. On the 10-volt range the resistance of the meter is 20,000 $\Omega \times 10 = 200,000 \ \Omega$ or 200 KΩ. Here, the equivalent resistance of R_m and R_2 in parallel is:

$$R_A = \frac{R_m \times R_2}{R_m + R_2}$$

$$R_A = \frac{200,000 \times 10,000}{200,000 + 10,000}$$

$$R_A = \frac{2,000,000,000}{210,000}$$

$$R_A = 9524\Omega \text{ or about } 9.52K\Omega$$

Notice that R_A is very close to the value of R_2. Therefore, the circuit operation is only slightly upset. The current increases only slightly to about:

$$I = \frac{E}{R} = \frac{6V}{10 \ K\Omega + 9.52 \ K\Omega} = \frac{6V}{19,520 \ \Omega} = 0.000307A \text{ or } 307 \ \mu A$$

The voltage across R_1 rises to about:

$$E_{R1} = I \times R_1$$
$$E_{R1} = 0.000307A \times 10,000 \ \Omega$$
$$E_{R1} = 3.07 \text{ volts}$$

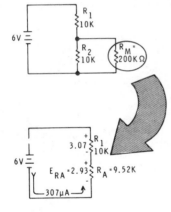

Figure 6-15 The 20,000 ohm-per-volt meter gives a much more accurate reading.

Meanwhile, the voltage across R_2 decreases slightly to:

$$E_{RA} = I \times R_A$$
$$E_{RA} = 0.000307A \times 9,520 \ \Omega$$
$$E_{RA} = 2.93V$$

Thus, instead of measuring 3 volts, the meter measures 2.93 volts. The inaccuracy is so small that it probably would never be noticed. The loading effect is minimized by using a voltmeter whose resistance is much higher than the resistance across which a voltage is to be measured.

THE OHMMETER

The basic meter movement can be used to measure resistance. The resulting circuit is called an ohmmeter. In its most basic form, the ohmmeter is nothing more than a meter movement, a battery, and a series resistance.

Basic Circuit

Figure 6-16 shows the basic circuit of the ohmmeter. The idea behind the ohmmeter is to force a current to flow through an unknown resistance then measure the current. For a given voltage, the current will be determined by the unknown resistance. That is, the amount of current measured by the meter is an indication of the unknown resistance. Thus, the scale of the meter movement can be marked off in ohms.

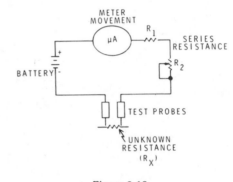

Figure 6-16
The basic ohmmeter.

The purpose of the battery is to force current through the unknown resistance. The meter movement is used to measure the resulting current. The test probes have long leads and they simplify the job of connecting the ohmmeter to the unknown resistor (R_x). Fixed resistor R_1 limits the current through the meter to a safe level. Variable resistor R_2 is called the ZERO OHMS adjustment. Its purpose is to compensate for battery aging.

Scale Calibration

In ohmmeters of this type, 0 ohms appears on the right side of the scale (at full scale deflection). The reason for this is shown in Figure 6-17A. Here the two test probes are shorted together. Thus, the unknown resistance (R_X) between the probes is equal to 0 ohms. In this case, the meter should deflect full scale to the 0 ohms marking of the scale. Full scale deflection for this meter is 50 μA. In order for the 1.5 volt battery to force 50 μA of current through the circuit, the total circuit resistance must be:

$$R_T = \frac{E}{I}$$

$$R_T = \frac{1.5V}{.00005A}$$

$$R_T = 30,000\Omega$$

The meter provides 2000 ohms while R_1 provides 22,000 ohms. Thus, R_2 must be set to exactly 6,000 ohms to insure a current of exactly 50 μA.

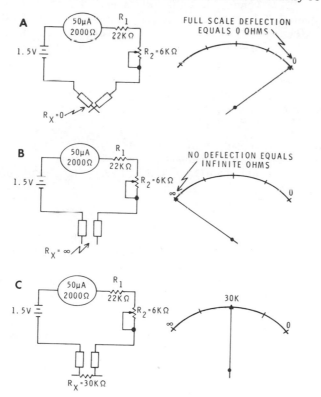

Figure 6-17 Calibrating the ends and center of the scale.

You may wonder why R_2 isn't a fixed 6,000 ohm resistor or better yet why R_1 isn't a fixed 28,000 ohm resistor. The reason for this is that the battery voltage will change as the battery discharges. If the battery voltage drops to 1.45 volts, then to achieve full scale deflection, the circuit resistance must be reset to:

$$R_T = \frac{E}{I}$$

$$R_T = \frac{1.45V}{.00005A}$$

$$R_T = 29,000\Omega$$

In this case R_2 would have to be reset to 5000 ohms to compensate for the lower voltage. R_2 is called the ZERO OHMS adjust. Adjusting this resistance to "zero the ohmmeter" is the first step in every resistance measurement.

We have seen that full scale deflection corresponds to an unknown resistance R_X of 0 ohms. Thus, the scale is marked 0 at this point. Now, what about the left side of the scale (no deflection)? Figure 6-17B illustrates this condition. Here an open circuit exists between the two test probes. This corresponds to an infinite resistance. No current flows through the meter movement and the pointer rests at the left side of the scale. Consequently, this point on the scale is marked with infinity symbol (∞). Thus, we have a scale with 0 ohms on the right and infinite ohms on the left.

Now let's see what resistance is represented by 1/2 scale deflection. The pointer will deflect to the center of the scale when the current is exactly 25 μA. This amount of current is caused by a total resistance of:

$$R_T = \frac{E}{I} = \frac{1.5V}{25\mu A} = \frac{1.5V}{.000025A} = 60,000\Omega$$

Since the meter, R_1, and R_2 have a combined resistance of only 30,000 ohms, the unknown resistance R_X supplies the other 30,000 ohms. That is, the meter deflects to half scale (25μA) when the unknown resistance has a value of 30KΩ. Consequently, the 1/2 scale point is marked 30K as shown in Figure 6-17C.

Using this same procedure, we can determine the amount of meter deflection for any value of R_x. Figure 6-18 illustrates that 1/3 full scale indicates an R_x of 60KΩ while 2/3 full scale indicates an R_x of 15KΩ. You can verify this by proving that 1/3 of 50μA or 16.66μA flows in the circuit shown in Figure 6-18A. Also, verify that 2/3 of 50μA or 33.33μA flows in the circuit shown in Figure 6-18B.

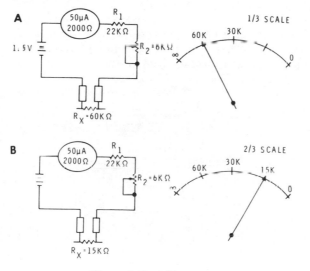

Figure 6-18   Calibrating for
1/3 and 2/3 scale deflection.

If enough points on the scale are found, the scale will take the form shown in Figure 6-19. There are two important differences between this scale and the ones used for current and voltage. First, the ohms scale is reversed with 0 on the right. Second, the scale is non-linear. For example, the entire upper half of the scale is devoted to a range of only 30KΩ; that is, from 0 to 30KΩ. However, notice that the second 30 kilohms (from 30KΩ to 60KΩ) takes up less than ¼ of the scale. The markings are squeezed closer together on the left side of the scale. On ammeters and voltmeters the scale is linear. That is, the scale is marked off in equal increments of current or voltage.

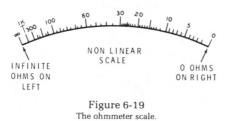

Figure 6-19
The ohmmeter scale.

Creating Higher Ranges

A one range ohmmeter would be of limited use. For this reason, multi-range ohmmeters have been developed. Two techniques have evolved for creating additional ranges. Both techniques are used in some ohmmeters.

Figure 6-20 shows how a higher resistance range can be implemented. First a switch is added to switch between the two ranges. Second a higher voltage battery is added. Finally, a higher value series resistor is required. To increase the range by a factor of 10, both the voltage and the total series resistance must be increased by a factor of 10.

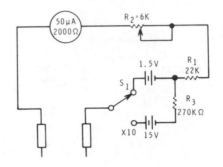

Figure 6-20   Creating higher resistance ranges.

When S_1 is in the position shown, the meter operates exactly like the one shown earlier in Figure 6-17 and 6-18. However, when S_1 is switched to the × 10 position, the 15 volt battery is switched in series with R_3, R_1, R_2, and the meter. The higher voltage will not cause excessive current through the meter since the series resistance has been increased by the addition of R_3. Notice that the total resistance in the circuit is now 300 KΩ. Thus, when the leads are shorted together, the current is still:

$$I = \frac{E}{R} = \frac{15V}{300K\Omega} = 50\mu A$$

The right side of the scale still represents 0 ohms. However, half-scale deflection (25μA) now occurs when the total resistance is:

$$R_T = \frac{15V}{25\mu A} = 600K\Omega$$

Of this, the meter, R_1, R_2, and R_3 supply 300KΩ. Therefore, the unknown resistance R_X must be 300KΩ. This means that the center of the ohmmeter scale now represents 300KΩ instead of 30KΩ. The range has been increased by a factor of 10.

202

Of course, this technique cannot be carried much further because increasing the range by an additional factor of 10 would require a 150 volt battery. Fortunately, the range described above is sufficient for general purpose use. It can measure resistance values up to several megohms. Resistors larger than this are rarely used in most electronic applications.

Creating Lower Ranges

The basic ohmmeter can also be modified to measure lower values of resistance. This is done by switching a small value shunt resistor in parallel with the meter and its series resistance.

Refer to Figure 6-21. With switch S_1 in the position shown, the ohmmeter operates exactly like the one shown earlier in Figure 6-17. However, when the position of S_1 is changed, R_3 is connected in parallel with the series combination of the meter, R_1 and R_2. The value of R_3 is 300 Ω or 1 percent of the combined series resistance of R_1, R_2 and the meter (30,000 Ω). Therefore, 99 percent of the current will flow through R_3 and only 1 percent will flow through the meter circuit. Recall that 25 μA of current is required for half-scale deflection of the meter. Let's determine what value of R_X will cause this amount of current to flow through the meter.

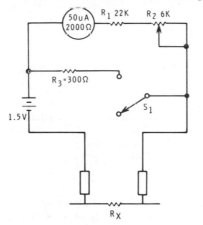

Figure 6-21 Creating lower resistance ranges.

The resistance of the meter circuit is now about 300 ohms. If an unknown resistance R_X of 300 ohms is now connected between the probes, the current from the battery becomes:

$$I = \frac{1.5V}{600\Omega} = 2.5mA$$

However, 99% of this current (2.475mA) flows through R_3. Only 1% or 25μA flows through the meter movement. Thus, half-scale deflection now represents an unknown resistance of 300 ohms instead of 30KΩ. Using the technique, lower ohmmeter ranges can be created.

Shunt Ohmmeter

The ohmmeters discussed up to this point are called *series* ohmmeters because the unknown resistance is always placed in series with the meter movement. A series ohmmeter can be recognized by its "backwards" scale. That is, 0 ohms is on the right while infinite ohms is on the left.

Another type of ohmmeter is called a shunt ohmmeter. Figure 6-22 illustrates the basic circuit of the shunt ohmmeter. This instrument gets its name from the fact that the unknown resistance is placed in parallel (shunt) with the meter movement. This completely changes the characteristics of the ohmmeter. For example, notice that when an open (infinite ohms) exists between the probes, 50μA of current flows through the meter movement. This produces full scale deflection. Consequently, infinite ohms is on the right side of the scale or at full scale deflection. Notice that this is just the opposite of the series ohmmeter.

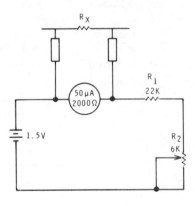

Figure 6-22
The shunt ohmmeter.

When the probes on the shunt ohmmeter are shorted together (representing an R_x of 0 ohms), the meter movement is shorted out. This produces no deflection. Thus, 0 ohms is on the left.

Recall that with the series ohmmeter, the half-scale reading was 30,000 ohms for the 50μA 2000Ω meter movement. However, with the shunt ohmmeter, this too is different. Here 25μA of current flows through the meter movement when R_x is the same resistance as the meter movement. Thus, half-scale deflection on the shunt ohmmeter is marked 2000 ohms.

The shunt ohmmeter has some disadvantages. For one thing, the battery discharges any time the ohmmeter is turned on. This is not the case with the series ohmmeter. It draws current from the battery only when a resistance is being measured.

Also, the meter movement in the shunt ohmmeter is more easily dammaged if the meter is inadvertently connected across a voltage source. In the series meter, the 28,000 ohms in series with the meter movement tends to limit the current. Even so, we should never connect either type of ohmmeter to a *live* circuit.

Finally, because the half-scale reading of the shunt ohmmeter is much less than that of the series ohmmeter, it is more difficult to accurately measure high resistance values on the shunt meter. However, it is easier to read low resistance values on the shunt meter for the same reason.

MULTIMETERS

Generally, the voltmeter, current meter, and ohmmeter are combined in a single instrument called a multimeter. There are several reasons for this. First it is much easier to carry a single instrument than it is to carry three separate meters. Second, because only a single meter movement, case, set of test leads, etc., are required, it is much cheaper to buy a single multimeter than to buy three separate single-function meters.

Basic Circuit

A schematic diagram of a very basic multimeter is shown in Figure 6-23. This meter has three dc voltage ranges, two ohmmeter ranges, and three current ranges. Function switch S_2 determines if the multimeter is to act as an ammeter, a voltmeter, or an ohmmeter. Range switch S_1 determines the range of the meter.

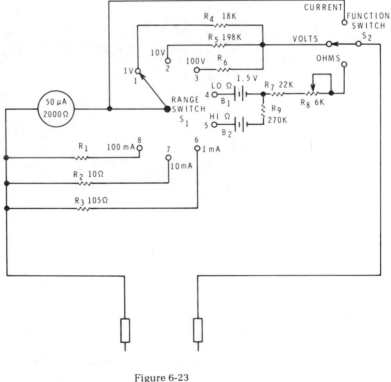

Figure 6-23
The multimeter.

When used as a voltmeter, S_2 must be placed in the VOLTS position and S_1 must be placed in one of the voltage positions (1V, 10V, or 100V). As the circuit is drawn, the multimeter is set up to measure dc voltages up to 1 volt. Compare the voltmeter section of this meter with the multirange voltmeter shown earlier in Figure 6-11. Notice that the two voltmeters are virtually identical.

The ohmmeter portion of the multimeter has only two ranges. On the LOΩ range, B_1 supplies the current which causes the meter to deflect when S_2 is in the OHMS position and a resistance is connected between the two test probes. In the HIΩ position a higher voltage battery and a larger value series resistor are used to increase the resistance range. The ohmmeter portion of the multimeter is exactly like the two range ohmmeter shown earlier in Figure 6-20.

When function switch S_2 is placed in the current position, the two test probes are connected directly to opposite ends of the meter movement. When the range switch S_1 is placed on one of the current ranges, a resistor is placed in parallel (shunt) with the meter movement. On the 1mA range, $50\mu A$ must flow through the meter movement when 1mA flows through the circuit under test. Thus, $950\mu A$ must flow through the shunt R_3. Since 19 times as much current flows through R_3 as through the meter movement, R_3's resistance must be 1/19th that of the meter movement.

$$R_3 = \frac{2000\,\Omega}{19} = 105\,\Omega$$

Using the same line of reasoning, the value of R_2 must be about 10 ohms.

Electronic Multimeter

All of the meters discussed up to this point are classified as electrical meters. They consist of a meter movement, precision resistors, and a battery for the ohmmeter. However, there is another family of meters called electronic meters. These contain electronic circuits which can amplify small voltages and currents. These instruments use devices such as transistors and vacuum-tubes which we have not yet discussed. For this reason, their circuitry will not be discussed in this course. Generally, the electronic meter is used in the same way as the electrical meter. However, one important difference is that the electronic meter has a much higher resistance than the electrical meter. For example, a good electrical meter has a sensitivity of 20,000 ohms per volt. Thus, on the 10 volt range, its resistance is 20,000 $\Omega \times 10 = 200$ KΩ. By contrast, most electronic meters have a resistance of 10 or 11 megohms on all dc ranges. Consequently, the electronic multimeter has very little loading effect on most circuits.

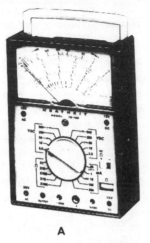 

A   B

It is not always easy to tell an electrical
meter (A) from the electronic type (B).
Courtesy Heath Company.

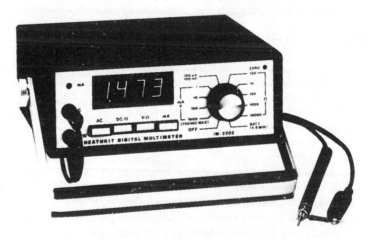

The digital voltmeter is easily recognized by its unique readout. Courtesy
Heath Company.

Measurement Inaccuracies

When using a multimeter, there are several ways in which inaccuracies can slip into our measurements. Some of the causes of inaccurate measurements have already been discussed. Voltmeter loading is a good example. This error is minimized by using a voltmeter which has a high resistance compared to the circuit resistances.

In somewhat the same way, the ammeter introduces an error when it is used to measure current. Since the ammeter has a finite value of resistance, it increases the overall circuit resistance when it is connected in series with the circuit. Of course, this reduces the current flowing in the circuit. As a result, the ammeter reads a value of current lower than the actual value without the meter. To minimize this error, an ammeter with a very low value of resistance must be used. The lower the ammeter resistance with respect to the circuit resistance, the smaller the error will be.

If the resistance value of the ammeter is known, its loading effect can be taken into consideration and a more accurate interpretation of the reading can be made. Often the resistance of the ammeter is given in the manufacturer's literature. If this value is not given the value can be determined using the procedure shown in Figure 6-24. First, the ammeter is set to the desired scale. Next, it is connected in series with a variable resistor and a voltage source as shown in Figure 6-24. R_1 is then adjusted for full scale deflection of the meter. Variable resistor R_2 is now added to the circuit as shown in Figure 6-24B. R_2 is adjusted until the meter reads exactly one half scale deflection. At this time half the current flows through R_2 and half flows through the meter. Thus, the resistance of R_2 must be equal to the resistance of the meter. Consequently, we can determine the resistance of the meter by removing R_2 and measuring the value of R_2 with the ohmmeter section of the multimeter. You may wonder why we do not measure the resistance of the ammeter directly with an ohmmeter. On high current ranges, this can be done if a separate ohmmeter is available. However, when the ammeter is set to a low current range, the ohmmeter may produce enough current to harm the ammeter. Earlier, we saw that the resistance of a meter movement is an important factor in designing meter circuits. The technique outlined above can be used to determine the resistance of any meter movement as well as that of an ammeter.

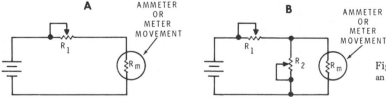

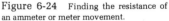

Figure 6-24   Finding the resistance of an ammeter or meter movement.

Aside from loading errors, the prime source of measurement inaccuracies is the characteristics of the meter itself. The meter movement may have an inaccuracy of ± 2% or ± 3% of full scale. Also, multiplier and shunt resistances may have a tolerance of 1%. Adding these inaccuracies, a typical multimeter may have an overall accuracy of ± 3% to ± 4% for dc voltage and current ranges.

Another error which can creep into measurements is caused by improperly reading the meter. The most common error is that caused by *parallax*. Figure 6-25 illustrates what is meant by parallax. In Figure 6-25A, the scale is viewed from directly in front of the meter. Viewing "straight on" like this, the meter indicates exactly 5 volts. If we move slightly to the right of the meter, the needle appears to point to the left of 5 volts as shown in Figure 6-25B. Also, if viewed from the left of center, the needle appears to point to the right of 5 volts as shown in Figure 6-25C. Obviously then, for a correct reading, we should always read the scale "straight on" from directly in front of the meter. However, because we have two eyes, one eye will be to the right of center while the other will be to the left of center. To prevent errors in reading the meter, we should close one eye, and read the scale "straight on" with the other eye. Some meters have a mirror on the scale to help eliminate parallax errors. With one eye closed, the needle is aligned with the reflection of the needle in the mirror. This ensures that the open eye is placed directly in front of the needle.

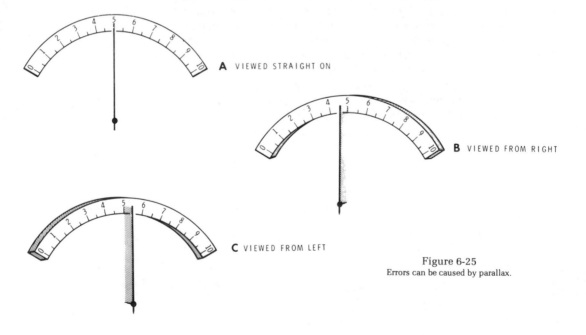

A VIEWED STRAIGHT ON

B VIEWED FROM RIGHT

C VIEWED FROM LEFT

Figure 6-25
Errors can be caused by parallax.

SUMMARY

The most popular type of meter movement is the d'Arsonval. It has a permanent magnet and a moving-coil. When current flows through the coil, a magnetic field is developed which interacts with the permanent magnet's field. The coil rotates as described by the right-hand motor rule. The amount of rotation is proportional to the current.

The main parts of the d'Arsonval meter movement are the moving-coil, the permanent magnet, the scale, pivots, bearings, springs, retaining pins, counter weights, and a zero adjust mechanism.

A variation of the d'Arsonval movement is the taut-band movement. In this device, thin metal bands replace the pivots, jeweled bearings, and spiral springs. This makes a more sensitive and more rugged movement.

A multirange ammeter can be constructed using a sensitive meter movement. A $50 \mu A$ movement can be used to measure a much higher current value by connecting a small value resistor (called a shunt) in parallel with the movement.

To determine the proper value of shunt, we must know the resistance of the meter movement. If the resistance of the meter movement is 1000 ohms, the current range can be doubled by connecting a 1000 ohm resistor in parallel. In the same way, if the shunt is to carry 9 times as much current as the meter movement, then the shunt resistance must be one-ninth that of the meter movement.

The accuracy of a meter movement is specified as a percentage of error at full scale deflection. If a 100mA meter has an accuracy of $\pm$ 3%, then the reading may be off by as much as $\pm$ 3mA at any point on the scale.

A current operated meter movement can be used to measure voltage. A 1mA, 1000Ω meter movement indicates full scale when the applied voltage is: $E = I \times R = 1 \text{ mA} \times 1 \text{ K}\Omega = 1V.$

To extend the voltage range, a voltage dropping resistor (called a multiplier) is connected in series with the meter movement. To extend the range of the above meter to 10V, the multiplier must drop 9V when the current is 1mA. Thus, its value must be:

$$R = \frac{E}{I} = \frac{9V}{1mA} = 9K\Omega.$$

The sensitivity of a meter movement is given in ohms per volt. It is determind by dividing the full-scale current rating into 1V. Thus, a 1mA meter has a sensitivity of:

$$\frac{1V}{1mA} = 1000 \text{ ohms per volt.}$$

The ohms per volt rating is important because it indicates the loading effect of a voltmeter. The higher the sensitivity, the less the loading effect will be.

The series ohmmeter consists of a battery, a meter movement, and a resistance all connected in series. Zero ohms is indicated by full scale deflection while infinite ohms are indicated by no deflection. The internal series resistance of the ohmmeter circuit must be such that it allows full scale current to flow when the test leads are shorted together. Thus, a 50 μA meter movement and a 4.5 volt battery require an internal resistance of:

$$R = \frac{E}{I} = \frac{4.5V}{50\mu A} = 90K\Omega.$$

A multimeter is a combination voltmeter, ohmmeter, and ammeter in a single case. A single meter movement is used. To measure voltage, multiplying resistors are switched in series. To measure current, shunt resistors are switched in parallel. To measure ohms, a battery and zero ohms adjust are switched in series with the meter movement.

Unit 7

DC CIRCUITS

INTRODUCTION

The purpose of this unit is to expand your knowledge of basic DC circuits. After a brief review of series and parallel circuits, you are introduced to some new applications of these circuits. Next you will study an entirely new circuit called a bridge. Finally, you will learn several new analytical tools for evaluating circuits.

SIMPLE CIRCUITS

Throughout your study of electronics, you will see certain circuits repeated over and over again. Some of the most used circuits are the easiest to understand. You have already studied the characteristics of several types of simple circuits. These included the series circuit, the parallel circuit, and the series-parallel circuit. Let's review the characteristics of these circuits. Then we will go on to some more advanced concepts.

Series Circuit

A series circuit is one in which the same current flows through all of the components in the circuit as shown in Figure 7-1. In such a circuit we may wish to find the current, the voltage dropped by any resistor, or the power dissipated by any resistor.

The current is the same in all the resistors. It is determined by dividing the total resistance into the applied voltage. The total resistance (R_T) in a series circuit is equal to the sum of the individual resistances. Thus, in Figure 7-1:

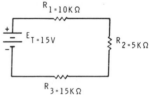

$$R_T = R_1 + R_2 + R_3$$

$$R_T = 10 \text{ K}\Omega + 5 \text{ K}\Omega + 15 \text{ K}\Omega$$

$$R_T = 30 \text{ K}\Omega$$

Figure 7-1
The Series Circuit

Once the total resistance is known, the current can be determined:

$$I = \frac{E}{R_T}$$

$$I = \frac{15\text{V}}{30 \text{ K}\Omega}$$

$$I = \frac{15\text{V}}{30,000\Omega}$$

$$I = 0.0005\text{A} = 0.5 \text{ mA}$$

Since the current is the same in all resistors, the voltage drop across any one resistor can be determined by multiplying the current times the value of that resistor. For example, the voltage across R_1 is:

$$E_{R1} = \quad I \times R_1$$

$$E_{R1} = \quad 0.0005A \times 10,000\Omega$$

$$E_{R1} = \quad 5 \text{ volts}$$

In the same way E_{R2} is:

$$E_{R2} = \quad I \times R_2$$

$$E_{R2} = \quad 0.0005A \times 5000\Omega$$

$$E_{R2} = \quad 2.5 \text{ volts}$$

Also, E_{R3} is:

$$E_{R3} = \quad I \times R_3$$

$$E_{R3} = \quad 0.0005A \times 15,000\Omega$$

$$E_{R3} = \quad 7.5 \text{ volts}$$

The voltage drops across the three resistors are worked out so that another important characteristic of the series circuit can be illustrated. The applied voltage E_T is equal to the sum of the voltage drops. That is:

$$E_T = \quad E_{R1} + E_{R2} + E_{R3}$$

$$E_T = \quad 5V + 2.5V + 7.5V$$

$$E_T = \quad 15V$$

Finally, the power dissipated by any resistor is equal to the current times the voltage drop across the resistor. Thus, the power dissipated by R_1 is:

$$P_{R1} = \quad I \times E_{R1}$$

$$P_{R1} = \quad 0.0005A \times 5 \text{ volts}$$

$$P_{R1} = 0.0025W \text{ or } 2.5mW$$

Review the above equations until you can compute any of the electrical quantities in the circuit.

Parallel Circuit

A simple parallel circuit is shown in Figure 7-2. Let's review the characteristics of this circuit. In a parallel circuit, the same voltage is applied to each branch. Thus, in Figure 7-2 the voltage is the same across each resistor:

$$E_T = E_{R1} = E_{R2} = E_{R3} = 15 \text{ volts}$$

That is, 15 volts is dropped across each resistor.

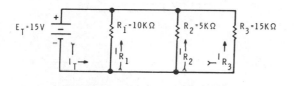

Figure 7-2
The parallel circuit

To find the current through any resistor, we divide the voltage by the resistance value of that resistor. For example, the current through R_1 is:

$$I_{R1} = \frac{E}{R_1}$$

$$I_{R1} = \frac{15V}{10 \text{ K}\Omega}$$

$$I_{R1} = \frac{15V}{10,000\Omega}$$

$$I_{R1} = 0.0015A = 1.5 \text{ mA}$$

Also, I_{R2} is:

$$I_{R2} = \frac{E}{R_2} = \frac{15V}{5,000\Omega} = 3 \text{ mA}$$

And I_{R3} is:

$$I_{R3} = \frac{E}{R_3} = \frac{15V}{15,000\Omega} = 1 \text{ mA}$$

Now, since the total current is the sum of the branch currents:

$$I_T = I_{R1} + I_{R2} + I_{R3}$$

$$I_T = 1.5 \text{ mA} + 3 \text{ mA} + 1 \text{ mA} = 5.5 \text{ mA}$$

We can determine the total current in another way by dividing the total resistance (R_T) into the applied voltage. To find the total resistance, we must use the equation for parallel circuits:

$$R_T = \cfrac{1}{\cfrac{1}{R_1} + \cfrac{1}{R_2} + \cfrac{1}{R_3}}$$

$$R_T = \cfrac{1}{\cfrac{1}{10,000\Omega} + \cfrac{1}{5,000\Omega} + \cfrac{1}{15,000\Omega}}$$

$$R_T = \cfrac{1}{.0001 + .0002 + .00006667}$$

$$R_T = \cfrac{1}{.00036667}$$

$$R_T = 2727 \text{ ohms}$$

Using this value of R_T, we find that the total current is:

$$I_T = \frac{E}{R_T} = \frac{15V}{2727\Omega} = 0.0055A = 5.5 \text{ mA}$$

Notice that this agrees with the value computed by adding the individual branch currents.

The power dissipated in any resistor can be found by multiplying the current through the resistor by the voltage dropped by the resistor. For example, the power dissipated by R_1 is:

$$P_{R1} = I_{R1} \times E_{R1} = 0.0015A \times 15V = 0.0225W = 22.5 \text{ mW}$$

Series-Parallel Circuit

Often a circuit will have both series and parallel current paths as shown in Figure 7-3A. To compute electrical quantities in this type of circuit, we must first simplify the circuit by redrawing it as shown in Figure 7-3B.

Notice that the two parallel resistors (R_2 and R_3) are replaced with an equivalent resistance (R_A). The value of R_A is determined by the equation:

$$R_A = \frac{R_2 \times R_3}{R_2 + R_3}$$

$$R_A = \frac{2,000 \times 2,000}{2,000 + 2,000}$$

$$R_A = \frac{4,000,000}{4000}$$

$$R_A = 1000\Omega = 1\ K\Omega$$

The resulting circuit shown in Figure 7-3B can now be handled like any other series circuit.

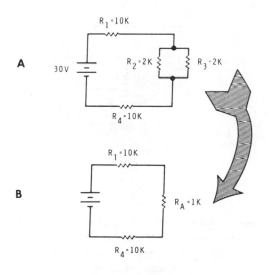

Figure 7-3
The series-parallel circuit

VOLTAGE DIVIDERS

One of the most useful series-parallel circuits is the voltage divider. It is frequently used at the output of a power supply to provide a number of output voltages which are distributed to different circuits.

The design of a voltage divider is complicated by the current drawn by the load. If we could ignore the load current, the design of the voltage divider would be extremely simple. For example, let's suppose we have a 30 volt power supply (or battery pack) and that we wish to obtain voltages of +15 volts and +30 volts. Figure 7-4A shows a circuit which appears to perform this function. Since R_1 and R_2 are the same size, each drops one-half of the applied voltage. If a voltmeter is placed across R_1, it will read +15 volts. Thus, the top of R_1 is at +15 volts with respect to ground. Naturally, the top of R_2 is at +30 volts with respect to ground. Thus, the circuit appears to meet the requirements set above. However, if we examine the circuit more closely, we find that the above conditions exist only under the *no load* condition.

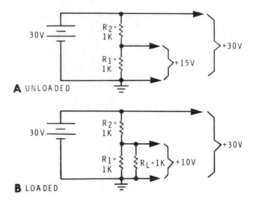

Figure 7-4
Simple Voltage Divider

When we attempt to use the +15 volts at the top of R_1 to drive a load, an interesting thing happens as shown in Figure 7-4B. Here a 1 KΩ load (R_L) is connected across the +15 volt supply. Notice that R_L is in parallel with R_1. Thus, the equivalent resistance of R_1 and R_L in parallel is:

$$R_T = \frac{R_1 \times R_L}{R_1 + R_L} = \frac{1000 \times 1000}{1000 + 1000} = \frac{1,000,000}{2000} = 500\Omega$$

The resistance from the top of R_1 to ground drops from 1 KΩ to 500Ω. R_2 is twice as large as the equivalent resistance. Consequently, R_2 drops 2/3 of the applied voltage while the equivalent resistance of R_1 and R_L drops 1/3. That is, R_2 drops 20 volts while R_1 and R_L in parallel drop only 10 volts. Thus, the voltage from the top of R_1 to ground decreases from +15V to +10V. When a load is connected across the divider, the divider no longer fulfills the requirements which were originally specified.

The point is, that in order to design a workable voltage divider, we must consider the current which flows through the load. To illustrate this point, let's assume we have three loads which must be driven from a single 12-volt power supply. Let's assume that the first load requires +12 volts at 1 ampere. Let's also assume that the second load requires 8 volts at 0.6 amperes while the third load requires 4 volts at 0.2 amperes. Figure 7-5 shows what the final circuit will look like.

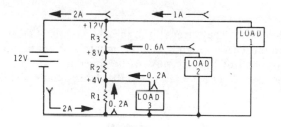

Figure 7-5
Designing the voltage divider.

Three resistors are used to develop the three desired voltages. The values of the three resistors are not given. In order to design the circuit, we must determine the values required for these resistors.

We have seen that the value of any resistor can be determined by Ohm's law if we know the current through the resistor and the voltage developed across it. If we analyze Figure 7-5 closely, we see that enough information is given to allow us to find the resistor values.

Look at R_1. The current through R_1 is given as 0.2 ampere. This is the only current in the circuit which is not flowing through one of the loads. This current is called the *bleeder* current and R_1 is called the bleeder resistor. Because the bleeder current does not flow through any of the loads, its value is not critical. It is generally selected to be about 10 percent of the total circuit current. For example, in Figure 7-5 the total current drawn from the supply is 2 amperes. Therefore, a bleeder current of 0.2 amperes is selected.

221

Once you determine what the bleeder current is to be, you can determine the value of R_1. The current through R_1 is 0.2A and the voltage across R_1 is 4V. Therefore:

$$R_1 = \frac{4V}{0.2A} = 20\Omega$$

The required value of R_2 can be determined by using a similar procedure. The current through R_2 is the sum of the 0.2A bleeder current through R_1 and the 0.2A through load 3. Therefore, the current through R_2 is 0.4A. The voltage at the bottom of R_2 is +4V while that at the top of R_2 is +8V. Consequently, the voltage across R_2 is 4 volts. Therefore:

$$R_2 = \frac{4V}{0.4A} = 10\Omega$$

Finally, look at R_3. The current through R_3 is the sum of the 0.6A through load 2 and the 0.4A through R_2. Thus, the total current through R_3 is 1A. The bottom end of R_3 is at +8V while the top of R_3 is at +12 volts. Therefore, the voltage across R_3 is 4 volts. Using Ohm's law, we find that the value of R_3 is:

$$R_3 = \frac{4V}{1A} = 4\Omega$$

Using this procedure, we can design a voltage divider for one or more loads. All we need to know is the source voltage plus the voltage and current requirements of each load.

To summarize, the step by step procedure is as follows:

1. Arbitrarily select a bleeder current which is about 10 percent of the total current in the circuit.

2. Using the bleeder current and the lowest voltage required by a load, compute the value of the bleeder resistor.

3. Using the total current through each resistor and the voltage dropped by the resistor determine the resistance values required.

A special form of voltage divider which is frequently used in electronics consists of a resistor in series with a load of some type. The resistor is called a *series dropping* resistor. Its purpose is to ensure that the load is operated at its proper voltage and current rating. For example, consider the problem of using a 5-volt relay in a system in which the only source of power is 12 volts. Let's assume that the relay is designed to operate at 100 mA. The solution is to connect a resistor in series with the relay coil so that the current is limited to 100 mA. Also, since the relay must operate at 5 volts, the series resistor must drop: 12 − 5 = 7 volts. Thus, the size of the resistor must be:

$$R = \frac{E}{I} = \frac{7V}{100 \text{ mA}} = 70 \text{ ohms.}$$

Dropping resistors are often found in series with relays, light bulbs, etc. The series dropping resistor and its load form a very simple voltage divider.

BRIDGE CIRCUITS

Another type of series-parallel circuit which is widely used is the bridge circuit. In its simplest form, the bridge circuit is made up of four resistors connected as shown in Figure 7-6A. The circuit has two input terminals and two output terminals. In DC applications, the input terminals are connected to some source of DC voltage such as a battery. Often a meter is connected across the output terminals as shown in Figure 7-6B.

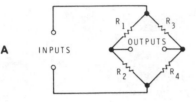
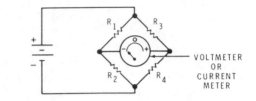

Figure 7-6
The bridge circuit

Balanced Bridge

Before we look at some of the ways that bridge circuits can be used, let's first investigate their characteristics. A bridge circuit may be either balanced or unbalanced. A balanced bridge circuit is one in which the voltage across the two output terminals is 0 volts. If the circuit shown in Figure 7-6B is balanced, a voltmeter or current meter connected across the output terminals (as shown) would read zero.

An example of a balanced bridge is shown in Figure 7-7A. Notice that R_1 has the same value as R_2. Thus, the voltage at point A with respect to ground must be one half the applied voltage or $+10$ volts. The voltage at point B is also $+10$ volts for the same reason. Therefore, point A is at the same potential as point B. A voltmeter connected from point A to point B will read 0 volts. Also, a current meter would read $0\mu A$ since no current can flow unless there is a difference in potential. Notice that the bridge is balanced when all four resistors have the same value.

Figure 7-7B shows that a balanced condition can exist even when all the resistors have different values. Here we can determine the voltage at point A if we know the current through R_2 (I_A).

$$I_A = \frac{E}{R_1 + R_2} = \frac{30V}{1\ K\Omega + 2\ K\Omega} = \frac{30V}{3\ K\Omega} = \frac{30V}{3000\Omega} = 0.01A$$

224

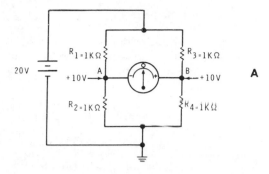 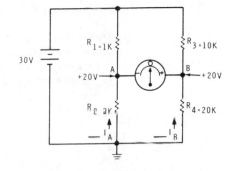

Figure 7-7
The balanced bridge

Now the voltage across R_2 can be determined:

$$E_{R2} = I_A \times R_2$$

$$E_{R2} = 0.01A \times 2000\Omega$$

$$E_{R2} = 20V$$

Thus, the voltage at point A with respect to ground is +20 volts.

Using a similar procedure, the voltage at point B can be determined. You will find that it is also +20V. Thus, there is no difference of potential between points A and B. Consequently, the bridge is balanced.

If you examine Figure 7-7B carefully, you will see that R_2 is twice the value of R_1 and that R_4 is twice the value of R_3. R_1 and R_2 form a voltage divider which determines the voltage at point A. R_3 and R_4 form a divider which determines the voltage at point B. As long as R_1 and R_2 are the same ratio as R_3 and R_4, the bridge will be balanced. Expressed as an equation, the bridge will be balanced when:

$$\frac{R_1}{R_2} = \frac{R_3}{R_4}$$

Notice that this equation is correct for both of the examples shown in Figure 7-7.

225

Unbalanced Bridge

Figure 7-8A shows a variation of the basic bridge circuit. Here R_4 is replaced with a potentiometer. If R_4 is set to the same value as the other three resistors (200Ω), the bridge is balanced. Both point A and point B are at +15 volts with respect to ground.

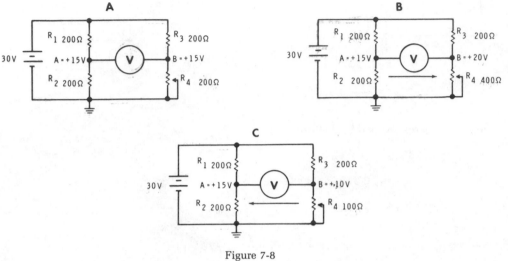

Figure 7-8
Unbalancing the bridge

The balanced condition can be upset by moving the arm of R_4. For example, in Figure 7-8B, R_4 is reset to 400Ω. The voltage at point A remains at +15 volts since neither R_1 nor R_2 are changed. However, the voltage at point B changes. Because R_4 is now twice as large as R_3, it drops twice as much voltage as R_3. Thus, R_4 now drops 20V while R_3 drops only 10V. Consequently, the voltage at point B is +20V with respect to ground. The bridge is no longer balanced because a difference of potential exists between points A and B. Because point B is more positive than point A, a minute current flows through the voltmeter as shown. A voltmeter connected in this way would read the 5 volt difference of potential.

Figure 7-8C shows that the bridge can also be upset in the opposite direction by making the value of R_4 smaller than the value of R_3. Here R_3 develops only 10 volts. Hence, the voltage at point B is only +10 volts. Thus, current will flow in the opposite direction through the meter.

Wheatstone Bridge

Now that you have an understanding of how the bridge circuit operates, let's see how it can be put to practical use. The first useful application of the bridge circuit was the *wheatstone bridge*. The wheatstone bridge is a device for measuring reistance.

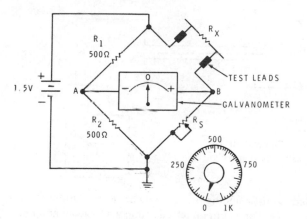

Figure 7-9
Wheatstone bridge

A simple wheatstone bridge is shown in Figure 7-9. R_1 and R_2 are fixed resistors that are carefully matched so that they are the same value. R_S is a 1 KΩ potentiometer which has a calibrated dial. At any setting, its value can be read directly from the dial. R_X is the unknown resistance which we wish to measure. It is connected between the two test leads of the bridge. The voltmeter has been replaced with a very sensitive current meter called a galvanometer. Unlike the current meters discussed earlier, the galvanometer can measure current flow in either direction. The center of the galvanometer scale is zero. Current flow in one direction is indicated by deflection to the left of center. Current flow in the opposite direction is indicated by deflection to the right of center.

To see how the bridge can be used to measure resistance, let's assume that a 210Ω resistor is connected across the test leads. Remember R_1 and R_2 form a voltage divider which determines the voltage at point A. Since R_1 and R_2 are equal, the bridge will be balanced only if R_X and R_S are also equal. Thus, R_S is set so that the galvanometer reads exactly zero. At this point the bridge is balanced so R_S must equal R_X. This value can be read directly from the calibrated dial of R_S. This gives us a simple method of finding the value of an unknown resistor.

This simplified bridge can measure resistances up to 1 KΩ. In actual bridges, R_S is replaced with a device called a decade resistor box. By setting rotary switches, any value of resistance from a fraction of an ohm to several megohms can be set. Thus, the bridge can be used to measure a wide range of resistances.

227

Self-Balancing Bridge

Another interesting application of a bridge circuit is shown in Figure 7-10. Here the four resistors are replaced by two potentiometers. Also, the meter is replaced by a small DC motor (M). The motor will turn anytime that the bridge is not balanced.

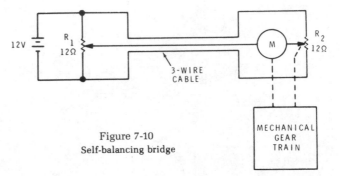

Figure 7-10
Self-balancing bridge

This circuit is called a self-balancing bridge and can be used for controlling motion at a distance. For example, let's assume we have an antenna on a roof which we wish to turn from ground level. The motor and R_2 are located on the roof with the antenna. R_1 and the power supply are located at ground level. The two circuits are connected by the 3-wire cable.

Let's assume that the circuit is initially balanced with the arms of both potentiometers set at the center position. Now if we wish to change the position of the antenna, we simply upset the balanced condition by moving the arm of R_1. This allows current to flow through the motor causing the motor to turn. The motor is connected through a gear train to the antenna. Thus, as the motor turns, the antenna turns also. The motor is also connected to the arm of potentiometer R_2. The motor turns until the arm of R_2 is at the same relative position as the arm of R_1. This rebalances the circuit and the motor stops turning. The circuit allows us to control the position of a remote antenna simply by setting the arm of a potentiometer.

Temperature Sensing Bridge

If one of the resistors in a bridge circuit is replaced with a thermistor, the bridge can be used to indicate temperature. Figure 7-11 shows a temperature sensing bridge.

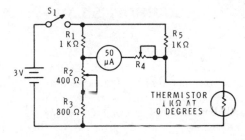

Figure 7-11
Temperature Sensing Bridge

R_1 and R_5 are 1 KΩ resistors. At 0 degrees the thermistor also has a resistance of 1 KΩ. If R_2 is set to 200 ohms then the fourth arm of the bridge also has a resistance of 1 KΩ. Consequently, the bridge is balanced and no current flows through the 50 μA meter movement. Thus, the 0 μA point on the meter scale can be labeled 0 degrees.

As the temperature to which the thermistor is exposed increases, the resistance of the thermistor decreases. This upsets the balanced condition and causes current to flow through the meter. The higher the temperature goes, the more current flows through the meter. Thus, the current is an indication of the temperature. The meter scale can be marked off in degrees rather than microamperes.

R_4 is included to provide a means of calibrating the high end of the scale. For example, let's assume we wish the upper limit of the scale to be 100°. The thermistor is exposed to a temperature of 100° and R_4 is adjusted for 50 μA of current through the meter movement. This causes full scale deflection. Thus, the point of full-scale deflection is labeled 100°.

Many series-parallel circuits can be analyzed using the techniques described earlier. However, more complex circuits cannot always be analyzed by such simple methods. Often a circuit will have several interconnected series-parallel branches and two or more voltage sources. Several techniques have been developed to help analyze circuits of this type. These techniques are generally grouped together under the name "network theorems."

A network is simply a circuit made up of several components such as resistors. Thus, the series-parallel circuits discussed earlier can be called networks. A network theorem is simply a logical method of analyzing a network. One of the most useful tools for analyzing a network is Kirchhoff's Law.

One form of Kirchhoff's law was discussed earlier. It states the relationship between the voltage rises and the voltage drops around a closed loop in a circuit. Recall that the sum of the voltage drops is equal to the sum of the voltage rises. This fact is referred to as Kirchhoff's voltage law.

The voltage law is deceptively simple. At this point it is almost self-evident. Nevertheless, this law is a powerful tool when used properly. Let's discuss it in more detail.

Kirchhoff's Voltage Law

The voltage law can be stated in several different ways. Up to now, it has been stated: *around a closed loop, the sum of the voltage drops is equal to the sum of the voltage rises.* Figure 7-12A illustrates this law.

Another way of saying the same thing is: *around a closed loop, the algebraic sum of all the voltages is zero.* It becomes apparent that this statement is true when we trace around the loop shown in Figure 7-12B. Notice that this is the same circuit shown in Figure 7-12A. To keep the polarity of the voltages correct, it is helpful to establish a rule for adding the voltages. A convenient 3-part rule is:

1. Choose which direction you prefer to trace. (Clockwise or counterclockwise works equally well.)
2. Trace around the circuit in the chosen direction. If the positive side of a voltage drop (or voltage rise) is encountered first, consider that voltage drop (or rise) to be positive.
3. If the negative side of a voltage drop (or voltage rise) is encountered first, consider that voltage drop (or rise) to be negative.

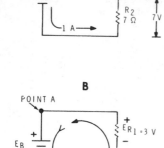

Figure 7-12
Kirchhoff's Voltage Law

For example starting at point A in Figure 7-12B, trace counterclockwise as shown recording each voltage encountered. The first voltage is E_B. Because the positive side of the battery is encountered first, the voltage is recorded as +10V. Next E_{R2} is encountered. This is recorded as −7V because we arrive at the minus side first. E_{R1} is recorded as −3V for the same reason. Thus, the sum of the voltages are:

$$+ 10 V - 7 V - 3V = 0V$$

Or stated as an equation:

$$E_B - E_{R2} - E_{R1} = 0V$$

As you can see, the sum of the voltage drops around the loop is zero.

In this example, we arbitrarily traced around the loop in the counterclockwise direction. However, tracing in the clockwise direction, we find that the sum is still zero.

Using Kirchhoff's Law

Look at the circuit shown in Figure 7-13A. We wish to find the current at all points in the circuit and the voltage dropped by each resistor. If we attempt to solve this problem using only Ohm's law, we run into insurmountable difficulties. However, if we use Kirchhoff's law with Ohm's law, the solution can be easily found.

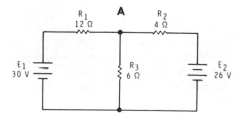

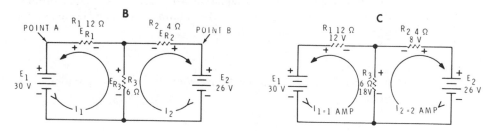

Figure 7-13
Using Kirchhoff's Law.

The first step in applying Kirchhoff's law is to assume a direction of current through each loop. It makes no difference if the assumed direction is incorrect. The algebraic sign of our solution tells us if the assumed direction is correct. For example, if our answer works out to $+3$ amps, the assumed direction is correct. If our answer works out to -3 amps, the assumed direction is incorrect. A minus quantity of current indicates that the assumed direction of current is reversed.

With the batteries connected as shown it appears that the loop currents will be in the direction shown by the arrow in Figure 7-13B. Let's assume that this is the direction of current flow.

The second step is to mark the polarity of the voltage drops across the resistors. We do this by marking negative polarity where current enters the resistor and positive polarity where current leaves the resistor. Using this procedure with the assumed direction of current flow, the polarity of voltage drops are marked as shown in Figure 7-13B.

The third step is to write an equation for each loop using Kirchhoff's voltage law. To do this, we start at a given point tracing around the loop in the assumed direction of current flow. We record each voltage rise and voltage drop following the procedure outlined earlier. Finally, we assume that the sum of all voltage drops and rises is equal to zero.

For example, let's trace out the loop through which I_1 flows. Let's start at the point labelled A in Figure 7-13B. Our equation becomes:

$$+ E_1 - E_{R3} - E_{R1} = 0$$

Notice that E_1 is assumed to be positive because its positive terminal is encountered first. In the same way, the voltage drops across R_3 and R_1 are assumed to be negative since their negative terminals are encountered first. The sum of the three terms are assumed to be equal to zero in accordance with Kirchhoff's law.

Using this same procedure an equation is developed for the loop through which I_2 flows. If we start at point B and follow the assumed direction of current flow, the equation becomes:

$$+ E_2 - E_{R3} - E_{R2} = 0$$

At this point we have two equations which describe the operation of the circuit. They are:

$$E_1 - E_{R3} - E_{R1} = 0$$

$$E_2 - E_{R3} - E_{R2} = 0$$

Notice that each equation contains a voltage rise and two voltage drops. The value of the voltage rises (E_1 and E_2) are given in Figure 7-13. Substituting 30 volts for E_1 and 26 volts for E_2 our equations become:

$$30\ V - E_{R3} - E_{R1} = 0$$

$$26\ V - E_{R3} - E_{R2} = 0$$

The next step is to manipulate these equations so that some of the other known values can be used. Let's start with the equation:

$$30\ V - E_{R3} - E_{R1} = 0$$

The term E_{R1} is the voltage drop across R_1. Since this voltage drop is caused by I_1:

$$E_{R1} = R_1 \times I_1 = R_1 I_1$$

Thus, the term $R_1 I_1$ can be substituted for E_{R1} in our equation. The equation becomes:

$$30V - E_{R3} - R_1 I_1 = 0$$

The term E_{R3} is the voltage drop across R_3. It is caused by currents I_1 and I_2. Thus:

$$E_{R3} = R_3 \times (I_1 + I_2) = R_3\ (I_1 + I_2)$$

Substituting, our equation becomes:

$$30V - R_3\ (I_1 + I_2) - R_1 I_1 = 0$$

Next, we substitute the values of R_3 and R_1:

$$30V - 6(I_1 + I_2) - 12\ I_1 = 0$$

We eliminate the parentheses by multiplying the -6 times each term within the parentheses:

$$30V - 6\ I_1 - 6\ I_2 - 12\ I_1 = 0$$

Combining $-6\ I_1$ and $-12\ I_1$ we have:

$$30V - 18\ I_1 - 6\ I_2 = 0$$

At this point, it is convenient to have the voltage rise on one side of the equation and the voltage drops on the other. We do this by subtracting 30V from both sides of the equation. This leaves:

$$- 18\ I_1 - 6\ I_2 = -\ 30V$$

233

Let's leave this equation for a moment and go back to the equation for the second loop. Recall that the equation is:

$$26V - E_{R3} - E_{R2} = 0$$

If we treat this equation the same as we did the first, the steps look like this:

$$26 \text{ V} - R_3 (I_1 + I_2) - R_2 I_2 = 0$$

Substitute values of R_3 and R_2:

$$26 \text{ V} - 6(I_1 + I_2) - 4 \text{ } I_2 = 0$$

Remove parentheses:

$$26 \text{ V} - 6 \text{ } I_1 - 6 \text{ } I_2 - 4 \text{ } I_2 = 0$$

Combine I_2 terms:

$$26 \text{ V} - 6 \text{ } I_1 - 10 \text{ } I_2 = 0$$

Subtract 26V from both sides:

$$- 6 \text{ } I_1 - 10 \text{ } I_2 = - 26V$$

Thus, our two original equations have been changed to:

$$- 18 \text{ } I_1 - 6 \text{ } I_2 = - 30V$$

$$- 6 \text{ } I_1 - 10 \text{ } I_2 = - 26V$$

Notice that each equation has two unknowns, I_1 and I_2. Equations of this type are called simultaneous equations. There are several methods of solving simultaneous equations. One of the simplest methods involves cancelling or eliminating one of the unknowns. This method works because of two simple facts:

1. We can do most anything we like to one side of an equation as long as we do the same thing to the other side. For example, we can add, subtract, multiply, or divide both sides of an equation by a number without upsetting the equality.
2. We can add two simultaneous equations together without upsetting the equality.

Consider our two simultaneous equations again:

$$- 18 \text{ } I_1 - 6 \text{ } I_2 = - 30 \text{ V}$$

$$- 6 I_1 - 10 \text{ } I_2 = - 26V$$

234

We can multiply both sides of the bottom equation by -3 so that our equations become:

$$- 18\ I_1 - 6\ I_2 = - 30V$$

$$+ 18\ I_1 + 30\ I_2 = + 78V$$

Notice that by multiplying the bottom equation by 3, we have changed the term containing I_1 in the bottom equation to the same magnitude but opposite sign as the term containing I_1 in the top equation. If the two equations are now added, the I_1 terms cancel:

$$- 18\ I_1 - 6\ I_2 = - 30V$$

$$\underline{+ 18\ I_1 + 30\ I_2 = + 78V}$$

$$0 + 24\ I_2 = + 48V$$

Thus, we end up with a single equation:

$$+ 24\ I_2 = + 48V$$

We can solve for I_2 by dividing both sides of the equation by 24:

$$I_2 = \frac{+48}{+24} = + 2 \text{ amps}$$

This positive value of current tells us that the assumed direction of current was correct. Now we can find I_1 by substituting $+2$ for I_2 in either of the above equations. That is:

$$- 18\ I_1 - 6\ I_2 = - 30V$$

Substitute $+2$ for I_2:

$$-18\ I_1 - 6\ (+2) = - 30V$$

$$- 18\ I_1 - 12 = - 30V$$

Add 12 to both sides:

$$- 18\ I_1 = - 18V$$

Divide by $- 18$:

$$I_1 = \frac{-18}{-18}$$

$$I_1 = +1 \text{ amp}$$

Thus, the currents in the circuit are as shown in Figure 7-15C. Once the currents are known, the voltage drops can be determined by Ohm's law. When computing the voltage drop across R_3 remember that both I_1 and I_2 flow through the resistor. Thus,

$$E_{R3} = R_3 (I_1 + I_2)$$

$$E_{R3} = 6\Omega (1A + 2A)$$

$$E_{R3} = 6\Omega (3A)$$

$$E_{R3} = 18V$$

Verify that the other voltage drops shown in Figure 7-15C are correct.

Kirchhoff's Current Law

Another form of Kirchhoff's Law involves current rather than voltage. Kirchhoff's current law can be stated in several different ways. One form states that in parallel circuits, the total current is equal to the sum of the branch currents. Stated another way, the current entering any point in a circuit is equal to the current leaving that same point. Figure 7-14 illustrates that this is simply two different ways of saying the same thing. Two branch currents of 1 ampere each are flowing. Thus, the total current is 2 amperes. Now look at point A. Notice that 2 amperes flow into this point. Consequently, 2 amperes must flow out of this point. One ampere flows up through R_1 while the other ampere flows over and up through R_2. Once again, an important law is simply common sense. Nevertheless, this simple law can be used in much the same way as the voltage law to evaluate networks. However, because Kirchhoff's voltage law is generally easier to use we will not be using the current law in this unit.

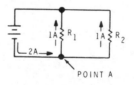

POINT A

Figure 7-14   The current
leaving point A is equal
to the current entering point A.

SUPERPOSITION THEOREM

The superposition theorem is the most logical of the network theorems that we will study. This theorem is widely used in physics, engineering, and even economics. It is used to analyze systems in which several forces are acting simultaneously to cause a net effect. It gives us a simple logical method of determining the net effect.

The principle behind the superposition theorem is simple. The net effect of several causes can be determined by finding the individual effect of each cause acting alone and then adding all the individual effects together. For example, suppose we have two batteries which are both forcing current to flow through a resistor. The net current can be determined by finding the individual currents caused by each battery and then adding the individual currents together.

In more formal terms, the superposition theorem states: *In a linear, bilateral network containing more than one voltage source, the current at any point is equal to the algebraic sum of the currents produced by each voltage source acting separately.* So that the terms *linear* and *bilateral* do not cause confusion, let's define these terms. A linear circuit is one in which the current is directly proportional to the voltage. If the voltage doubles, so does the current. Bilateral means that the circuit conducts equally well in either direction. The resistive networks which we have been studying are both linear and bilateral so they lend themselves to the superposition theorem.

The easiest way to see how this theorem works is to consider an example. Figure 7-15A shows a simple series circuit containing two voltage sources and two resistors. The 50-volt battery attempts to force current in a counterclockwise direction while the 75-volt battery attempts to force current in a clockwise direction. The problem is to find the net current in the circuit.

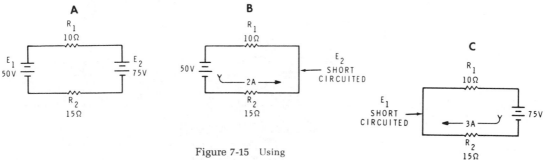

Figure 7-15 Using
the superposition theorem.

237

The first step is to consider only the current produced by the 50-volt battery. To do this, we mentally short circuit the 75-volt battery. This leaves the circuit shown in Figure 7-15B. Here the current is:

$$I = \frac{E}{R_T} = \frac{50V}{25\Omega} = 2 \text{ amperes.}$$

This current flows in the counterclockwise direction as shown.

Next we consider the current produced by the 75-volt battery. We mentally short circuit the 50-volt battery as shown in Figure 7-15C. The current is:

$$I = \frac{E}{R_T} = \frac{75V}{25\Omega} = 3 \text{ amperes.}$$

This current flows in the clockwise direction as shown.

As you can see, the 50-volt battery attempts to force 2 amperes in the counterclockwise direction. Simultaneously, the 75-volt battery attempts to force 3 amperes of current in the clockwise direction. Obviously then, the net current is 1 ampere in the clockwise direction.

Now let's consider a slightly more complex network. Figure 7-16A shows the circuit which we analyzed earlier using Kirchhoff's law. Let's use the superposition theorem to find the currents in the various parts of the circuit.

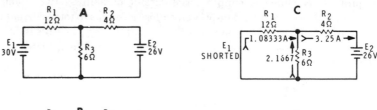

Figure 7-16 Applying
the superposition theorem
to a more complex network.

238

The first step is to redraw the circuit as shown in Figure 7-16B. Here we are interested only in the current produced by E_1. Therefore, E_2 is shorted. The total resistance in this circuit is:

$$R_T = R_1 + \frac{R_2 \times R_3}{R_2 + R_3}$$

$$R_T = 12 + \frac{4 \times 6}{4 + 6}$$

$$R_T = 12 + \frac{24}{10}$$

$$R_T = 12 + 2.4$$

$$R_T = 14.4\Omega$$

Thus, the total current produced by E_1 is:

$$I = \frac{E_1}{R_T} = \frac{30V}{14.4\Omega} = 2.08333 \text{ amperes.}$$

With this amount of current, the voltage drop across R_1 is:

$$E_{R1} = I \times R_1 = 2.08333A \times 12\Omega = 25V.$$

This leaves 5V across R_2 and R_3. Thus, the current through R_2 is:

$$I_{R2} = \frac{E_{R2}}{R_2} = \frac{5V}{4\Omega} = 1.25A.$$

And, the current through R_3 is:

$$I_{R3} = \frac{E_{R3}}{R_3} = \frac{5V}{6\Omega} = 0.8333A.$$

Thus, the currents produced by E_1 are as shown in Figure 7-16B.

Next, we consider the currents produced by E_2 with E_1 shorted. The circuit is shown in Figure 7-16C. Here the total resistance is:

$$R_T = R_2 + \frac{R_1 \times R_3}{R_1 + R_3}$$

$$R_T = 4 + \frac{12 \times 6}{12 + 6}$$

$$R_T = 4 + \frac{72}{18}$$

$$R_T = 4 + 4$$

$$R_T = 8\Omega$$

Thus, the total current is:

$$I_T = \frac{E_2}{R_T} = \frac{26V}{8\Omega} = 3.25A.$$

With this current, the voltage drop across R_2 is:

$$E_{R2} = I \times R_2 = 3.25A \times 4\Omega = 13V.$$

This leaves 13 volts across R_1 and R_3. Thus, the current through R_1 is:

$$I_{R1} = \frac{E_{R1}}{R_1} = \frac{13V}{12\Omega} = 1.08333A.$$

Also, the current through R_3 is:

$$I_{R3} = \frac{E_{R3}}{R_3} = \frac{13V}{6\Omega} = 2.1667A.$$

Thus, the currents produced by E_2 are as shown in Figure 7-16C.

The final step is to superimpose the two circuits. In Figure 7-16B, the current through R_1 is 2.0833A flowing to the left. In Figure 7-16C, the current through R_1 is 1.08333A flowing to the right. Obviously then, the net current is 1A to the left. Combining the other individual currents in the same way, we find the net currents are as shown in Figure 7-16D. Also, if you refer back to Figure 7-13C, you will see that these are the same values computed earlier using Kirchhoff's law.

240

THEVENIN'S THEOREM

Another important and powerful tool for simplifying and analyzing networks is Thevenin's theorem. Unlike the laws and theorems discussed earlier, the reason that Thevenin's theorem works is not at all obvious. Fortunately, we can use this theorem without understanding *why* it works.

This theorem allows us to replace any two-terminal network of voltage sources and resistors, no matter how they are interconnected, with a single voltage source in series with a single resistor. A two-terminal network is simply a network which has two output terminals. For example, two-terminal networks such as those shown in Figure 7-17A can be replaced by an equivalent circuit like that shown in Figure 7-17B.

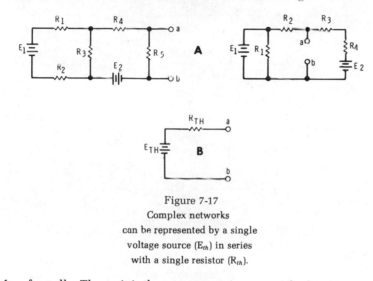

Figure 7-17
Complex networks
can be represented by a single
voltage source (E_{th}) in series
with a single resistor (R_{th}).

More formally, Thevenin's theorem states: *Any network of resistors and voltage sources, if viewed from any two terminals in the network, can be replaced with an equivalent voltage source (E_{th}) and an equivalent series resistance (R_{th}).* Of course, there are definite rules concerning the values assigned to E_{th} and R_{th}.

Perhaps the best way to understand this theorem is to consider an example. Figure 7-18A shows a simple circuit which can be easily analyzed using Ohm's law. In this example, R_3 is the load while the remaining circuitry is considered to be the source. Use Ohm's law to determine the current through R_3. Do not read further until you have calculated the value of I_{R3}. Now let's find the same current using Thevenin's theorem.

241

The first step is to mentally disconnect the load R_3 from the rest of the circuit as shown in Figure 7-22B. Notice that the circuit to the left of terminals a and b now has the same general form as those circuits shown in Figure 7-21A. Thus, this two-terminal network can be replaced with a "Thevenin equivalent" like the one shown earlier in Figure 7-21B. The only problem is to find the proper value of E_{th} and R_{th}.

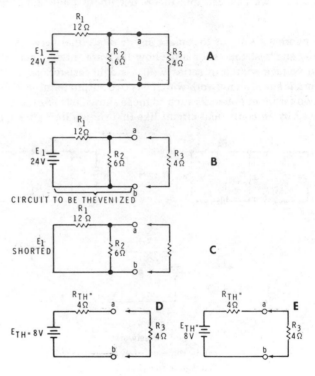

Figure 7-18
Using Thevenin's theorem

Let's find E_{th} first. E_{th} is the voltage at points a and b when R_3 is disconnected as shown in Figure 7-18B. This is called the open circuit voltage. Remember, it is the voltage between points a and b when nothing is connected to these points except the circuit that is to be "thevenized." Since E_1, R_1, and R_2 now form a simple series circuit, the voltage between points a and b can be easily computed. First, we find the current in the circuit:

$$I = \frac{E}{R_T} = \frac{24V}{18\Omega} = 1.333 \text{ amperes.}$$

242

Notice that the voltage between points a and b is the voltage drop across R_2 or E_{R2}. Thus, $E_{th} = E_{R2} = I \times R_2 = 1.333A \times 6\Omega = 8$ volts.

Now let's determine the value of R_{th}. R_{th} is the resistance between terminals a and b when the voltage source is shorted out. When E_1 is shorted as shown in Figure 7-18C, the resistance between a and b is:

$$R_{th} = \frac{R_1 \times R_2}{R_1 + R_2} = \frac{12 \times 6}{12 + 6} = \frac{72}{18} = 4\Omega.$$

Thus, the network shown in Figure 7-18B can be represented by the Thevenin equivalent shown in Figure 7-18D. Now let's reconnect R_3 between Terminals a and b as shown in Figure 7-18E. Notice that the series parallel network has been redrawn as a simple series equivalent circuit. The current through R_3 can now be easily determined:

$$I = \frac{E}{R_T} = \frac{8V}{8\Omega} = 1 \text{ ampere}.$$

The current through R_3 is 1 ampere. This should be the same value which you computed using Ohm's law.

You may wonder why Thevenin's theorem would be used at all on a simple circuit like that shown in Figure 7-18A. After all, we can find E_{R3} and I_{R3} using only Ohm's law. However, there are times when this theorem is extremely valuable even with simple circuits like this one. Suppose for example, that we had to compute E_{R3} and I_{R3} for 100 different values of R_3. Using only Ohm's law, we would have to compute 100 different series-parallel problems. Using Thevenin's theory we would still work one hundred problems. However, because the Thevenin equivalent would be the same for all problems, the problems would be the comparatively simple series-circuit type rather than the more complex series-parallel type.

This theorem also allows us to solve complex networks which cannot be analyzed using only Ohm's law. For example, Figure 7-19A shows our old familiar problem again. We have already analyzed this circuit using Kirchhoff's law and the superposition theorem. Now let's try Thevenin's theorem to see if we arrive at the same answers.

Let's find the current through R_3. We do this by thevenizing the entire circuit except for R_3. Thus, the first step is to remove R_3 from the circuit by opening points a and b. The resulting circuit is shown in Figure 7-19B. Notice that in this curcuit, E_1 tries to force current counterclockwise while E_2 tries to force current clockwise. The net voltage causing current flow is $E_1 - E_2 = 30V - 26V = 4V$. The total resistance is $R_1 + R_2 = 12\Omega + 4\Omega = 16\Omega$. Thus, the current is $4V \div 16\Omega = 0.25$ amperes in the counterclockwise direction. Verify that the voltage drops across R_1 and R_2 are the polarity and magnitude shown.

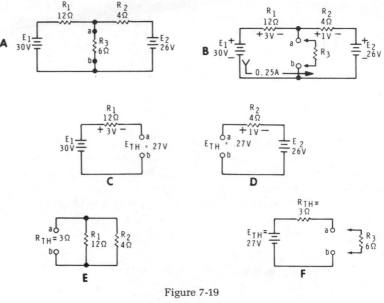

Figure 7-19

Using Thevenin's theorem

We can now determine E_{th}. Starting at points a and b and looking to the left, we see the circuit shown in Figure 7-19C. Notice that the voltage between a and b is 30V − 3V = 27V. You may wonder what value of E_{th} we would find if we look to the right of terminals a and b as shown in Figure 7-23D. The voltage is 1V + 26V = 27V. Notice that the value of E_{th} is the same because the two paths are in parallel.

Next, we find R_{th} by determining the resistance between points a and b when the two voltage sources are short circuited. As shown in Figure 7-19E, the resistance between points a and b is simply the value of R_1 and R_2 in parallel. Thus:

$$R_{th} = \frac{R_1 \times R_2}{R_1 + R_2} = \frac{12 \times 4}{12 + 4} = 3\Omega.$$

Thus, the Thevenin equivalent is the circuit shown in Figure 7-19F.

The final step is to reconnect R_3 to points a and b and then determine the current:

$$I = \frac{E}{R_T} = \frac{27V}{9\Omega} = 3 \text{ amperes}.$$

This agrees with our earlier findings for the current through R_3.

NORTON'S THEOREM

Using Thevenin's theorem, we found that a circuit like that in Figure 7-20A can be represented by the equivalent circuit in Figure 7-20B. Norton's theorem gives us a slightly different way of representing the same circuit. Thevenin's theorem uses an equivalent *voltage* source (E_{th}) in *series* with an equivalent resistance (R_{th}). On the other hand, Norton's theorem uses an equivalent *current* source (I_N) in *parallel* with an equivalent resistance (R_N). Thus, the circuit in Figure 7-20A can also be represented by the Norton equivalent shown in Figure 7-20C.

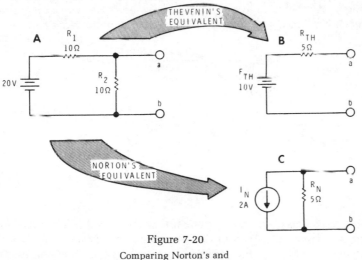

Figure 7-20
Comparing Norton's and
Thevenin's equivalent circuits.

Current Sources and Voltage Sources

At this point some words of explanation are in order. The idea of a voltage source is easy to visualize. An ideal voltage source is a device which produces the same output voltage regardless of the current drawn from it. As used in most electronic devices, a battery can be considered a nearly ideal voltage source. A large 12-volt battery will produce an output of approximately 12 volts whether the load current is 0 amperes, 1 ampere or even 10 amperes. The reason for this is that the resistance of the battery is very low compared to the load resistance connected to it. An ideal voltage source would have an internal resistance of 0 ohms. In most electronic devices the internal resistance of the battery or power supply is negligible compared to other circuit resistances. Thus, a battery can generally be considered as a nearly ideal voltage source. In Thevenin's theorem, E_{th} is considered to be an ideal voltage source having 0 ohms resistance.

245

The idea of a current source is similar. Whereas a voltage source has a certain voltage rating, the current source has a certain current rating. An ideal current source will deliver its rated current regardless of what value of resistance is connected across it. Figure 7-20 illustrates this point. The voltage source (E_{th}) in Figure 7-20B produces exactly 10 volts regardless of what resistance value is connected across terminals a and b. In the same way, the current source in Figure 7-20C produces 2 amperes regardless of the resistance value connected across terminals a and b.

A current source can be visualized as a voltage source with an enormously high internal resistance. Consider, for example, the circuit shown in Figure 7-21. Here a 10-volt battery is shown in series with a 1 megohm resistor. If points a and b are shorted, the current in the circuit is:

$$I = \frac{E}{R} = \frac{10V}{1,000,000\Omega} = 10 \ \mu A$$

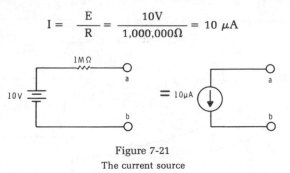

Figure 7-21
The current source

If a 10Ω resistor is placed across a and b, the current will still be 10 μA for all practical purposes. Even a 1 KΩ resistor would not cause a noticeable drop in current. Thus, this circuit acts as a 10 μA current source. Of course, this is not an ideal current source because if a large enough resistance is placed between point a and b the current will decrease. An ideal current source would have an infinite resistance and the current output would be constant regardless of the load resistance.

Notice that the symbol for the current source is a circle with an arrow. In this course, the arrow will point in the direction of electron flow through the current source.

Finding The Norton Equivalent

Norton's theorem allows us to represent a circuit containing voltage sources and resistors as a current source (I_N) in parallel with an equivalent resistance (R_N). Thus, although an ideal current source is impossible to build we can represent the most simple or the most complex circuit as if it were an ideal current source and a parallel resistor. As with Thevenin's theorem, there are strict rules that tell us what the value of I_N and R_N must be to represent a specific circuit.

The value of the current source (I_N) is equal to the current which flows through the terminals of the network when they are shorted. As an example, refer back to Figure 7-20A. The first step in converting this circuit to its Norton equivalent is to short terminals a and b to determine the short circuit current. Notice that doing this will short out R_2. Thus, the total resistance in the circuit is the 10 ohms of R_1. Consequently, the value of the current source is:

$$I_N = \frac{E}{R} = \frac{20V}{10\Omega} = 2 \text{ amperes.}$$

The size of the parallel resistor (R_N) is equal to the resistance between terminals a and b with the voltage sources shorted. Thus, in Figure 7-20A, the resistance is that of R_1 and R_2 in parallel of 5 ohms. Notice that the method for finding the resistor value is the same in both Thevenin's and Norton's theorems. The difference is that in the case of Norton's theorem the resistor is placed in parallel with the current source. Using these rules, the circuit in Figure 7-20A reduces to the Norton equivalent in Figure 7-20C.

With nothing connected between terminals a and b, the entire 2 amperes flow through R_N. However, if a load resistor is connected across terminals a and b, the current will split between R_N and the load resistor. The amount of current that each resistor conducts is inversely proportional to its resistance. If the load resistor has the same value as R_N, each will pass 1 ampere. If the load resistor is larger, it will pass proportionately less current.

Figure 7-22A is a repeat of the circuit which we have been evaluating with each of the network theorems. To illustrate Norton's theorem, let's use it to solve for the current through R_3. Since we are interested in the current through R_3, we wish to find a Norton equivalent for the circuit which connects to R_3.

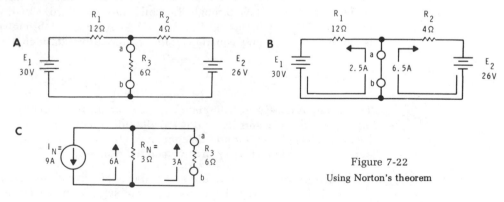

Figure 7-22

Using Norton's theorem

To find the current source (I_N) we must imagine a short circuit across terminals a and b as shown in Figure 7-22B. The current from b to a caused by E_1 is:

$$I = \frac{E_1}{R_1} = \frac{30V}{12\Omega} = 2.5A$$

Also, E_2 causes a current to flow from b to a:

$$I = \frac{E_2}{R_2} = \frac{26V}{4\Omega} = 6.5A$$

Thus, the total current from b to a is $2.5 + 6.5 = 9A$. This is the value of the current source (I_N) as shown in Figure 7-22C.

Next we find the shunt resistance (R_N) by mentally shorting E_1 and E_2 and measuring the resulting resistance. Obviously, the resistance is equal to R_1 and R_2 in parallel. Thus:

$$R_N = \frac{R_1 \times R_2}{R_1 + R_2}$$

$$R_N = \frac{12 \times 4}{12 + 4} = \frac{48}{16} = 3\Omega$$

Thus, the circuit shown in Figure 7-22A can be redrawn as shown in Figure 7-22C. Notice that everything except R_3 has been reduced to I_N and R_N.

Finally, we can find the current through R_3 by determining how the 9A from the current source is distributed between R_N and R_3. Obviously, R_N will draw more of the current since it is smaller. In fact, R_N will draw twice as much current since it is one-half the size of R_3. Thus, two-thirds of the current will flow through R_N while only one-third will flow through R_3. The current through R_N is $^2/_3 \times 9A = 6A$ while that through R_3 is $^1/_3 \times 9A = 3A$. A handy equation for finding the current through R_3 is:

$$I_{R3} = \frac{R_N}{R_N + R_3} \times I_N = \frac{3}{3 + 6} \times 9 = \frac{3}{\cancel{9}} \times \cancel{9} = 3A$$

This equation is used to determine how the current (I_N) is distributed between the two resistors $(R_N$ and $R_3)$. While this equation is especially useful when applying Norton's theorem, it can be used to determine the current distribution between any two parallel resistors.

Our procedure ends as we find that 3 amperes of current are flowing through R_3. Earlier, we solved this same problem in other ways and arrived at the same answer.

Norton-Thevenin Conversions

It has probably occurred to you that there are striking similarities between Norton's theorem and Thevenin's theorem. Norton's theorem represents a circuit as a current source and a shunt resistor, while Thevenin's theorem represents the same circuit as a voltage source in series with the same value resistor. Since a given circuit can be represented in either form, there must be some way of converting directly from a Norton equivalent to a Thevenin equivalent and vice versa.

Figure 7-23A shows a two-terminal network which can be represented as the Thevenin equivalent shown in Figure 7-23B or as the Norton equivalent in Figure 7-23C. We can convert the Thevenin form directly to the Norton form simply by applying Norton's theorem to the Thevenin equivalent circuit. If we short points a and b, the current will be:

$$I_N = \frac{E_{th}}{R_{th}} = \frac{6V}{4\Omega} = 1.5A$$

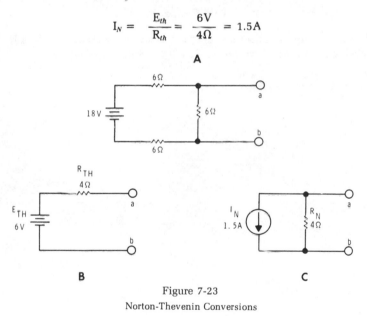

A

B   **C**

Figure 7-23

Norton-Thevenin Conversions

Next, we find R_N by shorting out E_{th} and measuring the resistance between points a and b. Obviously, R_N will be the same as R_{th}. Thus, the Norton equivalent shown in Figure 7-23C is derived.

Conversion in the opposite direction is just as easy. We simply apply Thevenin's theorem to the Norton equivalent. The open circuit voltage between a and b is:

$$E_{th} = I_N \times R_N = 1.5A \times 4\Omega = 6V$$

Now, to find R_{th}, we must know how to handle a current source. Recall that an ideal voltage source has 0 resistance and, thus, is replaced with a short when finding R_{th} or R_N. However, as we have seen, an ideal current source has infinite resistance. Thus, when finding R_{th} or R_N, a current source is considered an open circuit. Thus, if we open the current source, the resistance between points a and b is 4 ohms. As before: $R_{th} = R_N$.

To summarize, we can convert from the Thevenin form to the Norton form by applying two simple equations:

$$I_N = \frac{E_{th}}{R_{Th}} \text{ and } R_N = R_{th}$$

Likewise, we can convert from the Norton form to the Thevenin form by applying the equations:

$$E_{th} = I_N \times R_N \text{ and } R_{th} = R_N$$

The other thing we must remember is that in the Thevenin form R_{th} is in series while in the Norton form R_N is in shunt.

SUMMARY

In a series circuit the total resistance is found by adding the individual resistance values:

$$R_T = R_1 + R_2 + R_3 + \ldots$$

The current is the same at all points in the circuit and the sum of the voltage drops is equal to the applied voltage.

In parallel circuits, the total resistance is less than that of any branch. With two branches:

$$R_T = \frac{R_1 \times R_2}{R_1 + R_2}$$

With more than two branches:

$$R_T = \frac{1}{\dfrac{1}{R_1} + \dfrac{1}{R_2} + \dfrac{1}{R_3} + \ldots}$$

The voltage is the same across all branches and the total current is the sum of the branch currents.

Simple series-parallel circuits can usually be reduced to simpler forms by combining resistors using one or more of the above formulas.

A voltage divider is used to produce two or more output voltages from a common higher voltage.

A series dropping resistor is used to control the value of current and voltage applied to a load.

A bridge circuit generally consists of four resistances connected together so that there are two input terminals and two output terminals. Resistor values can be selected so that the bridge is balanced. In this condition, the voltage between the two output terminals is 0. The bridge can be used with a meter to measure resistance or temperature.

Kirchhoff's law provides a way of analyzing circuits which cannot be analyzed using Ohm's law alone. Kirchhoff's voltage law states that the sum of the voltages around a closed loop is 0. When a circuit has two or more loops, an equation can be written for each loop. The equations can then be combined in such a way that only one unknown is left. The unknown quantity can therefore be determined. Kirchhoff's current law states that the current entering a point is equal to the currents leaving that point. It too can be used to analyze complex circuits.

The superposition theorem gives us a logical way of analyzing circuits which have more than one voltage source. The effect of each voltage source is considered one at a time with all other voltage sources shorted out. The individual effects are then combined to determine the net effect of all voltage sources.

Thevenin's theorem is a handy tool for analyzing networks because it allows us to represent a complex two-terminal network as a single voltage source in series with a single resistor.

Norton's theorem is equally handy because it allows us to represent a complex two-terminal network as a single current source in parallel with a single resistor.

The maximum power transfer theorem states that maximum power is transferred to a load when the load resistance is equal to the resistance of the power supply.

Unit 8

INDUCTANCE
AND
CAPACITANCE

INTRODUCTION

Inductors and capacitors are generally considered to be AC devices. However, they do have some important DC characteristics which you should know about. This unit will serve as an introduction to these components.

INDUCTANCE

In an earlier unit on magnetism, you learned two rules that are quite important to the study of inductance. First, you saw that when current flows through a conductor, a magnetic field builds up around the conductor. Second, you saw that when a conductor is subjected to a moving magnetic field, a voltage is induced into the conductor. These two rules form the basis for a phenomenon called *self-induction*.

Self-Induction

Before we discuss self-induction, we must first differentiate between two different conditions which can exist in any DC circuit. The first is called the *steady-state condition* while the other is called the *transient condition*. Up to this point, we have considered only the steady-state condition. In most DC circuits, this condition is reached within a fraction of a second after power is applied. In this condition, the current in the circuit has reached the value computed by Ohm's law. However, the current does not reach the steady-state value instantaneously. There is a brief period called the *transient time* in which the current builds up to its steady-state value. Thus, the transient condition exists for an instant after power is initially applied to a circuit. In circuits containing only resistors, the transient condition exists for such a short period of time, that it can be detected only with sensitive instruments. However, if inductors or capacitors are used in the circuit, the transient condition may be extended so that it is readily apparent.

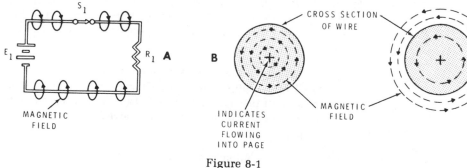

Figure 8-1
Self-induction

255

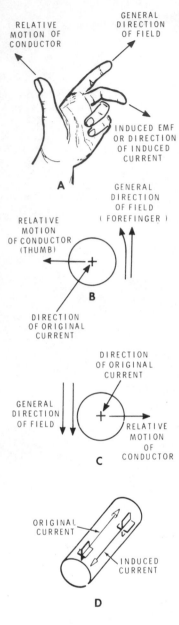

RELATIVE MOTION OF CONDUCTOR

GENERAL DIRECTION OF FIELD

INDUCED EMF OR DIRECTION OF INDUCED CURRENT

A

GENERAL DIRECTION OF FIELD (FOREFINGER)

RELATIVE MOTION OF CONDUCTOR (THUMB)

B

DIRECTION OF ORIGINAL CURRENT

DIRECTION OF ORIGINAL CURRENT

GENERAL DIRECTION OF FIELD

RELATIVE MOTION OF CONDUCTOR

C

ORIGINAL CURRENT

INDUCED CURRENT

D

Figure 8-2
Determining the
direction of the induced current

During the transient time when the current is changing from zero to some finite value, the phenomenon called self-induction occurs. Recall that a magnetic field builds up around any conductor when current flows. Also, when a conductor is cut by a moving magnetic field, a voltage is induced into the conductor. Keeping these two facts in mind, consider what happens during the transient time in the DC circuit shown in Figure 8-1. When S_1 is closed, current begins to flow and a magnetic field builds up around the conductor as shown. However, the magnetic field does not just suddenly appear, it must build up from the center of the wire. If we look at a cross section of the wire as shown in Figures 8-1B and C, we see that the magnetic field builds up over a period of time.

As the magnetic field expands from the center of the wire, it cuts the wire. This fulfills the requirements for inducing a voltage into the wire. Thus, the sequence of events is as follows:

1. the switch is closed
2. current begins to flow through the wire
3. a magnetic field begins to build up around the wire
4. the moving magnetic field induces a voltage into the wire

By using the left-hand generator rule discussed earlier, we can determine the polarity of the induced voltage and the direction of the induced current. Figure 8-2A shows the left-hand generator rule again to refresh your memory. Let's apply this rule to the cross section of the wire shown in Figure 8-1B. The tail of the arrow indicates that current is flowing into the page. Thus, the direction of the magnetic field is counterclockwise around the wire. Consequently, on the right side of the wire the general direction of the field is as shown in Figure 8-2B. Also, as the magnetic field expands outward on the right, the relative motion is the same as if the conductor had moved to the left. Apply the left-hand rule by pointing your thumb and forefinger as shown. Notice that your middle finger which indicates the direction of the induced current points out of the page. Thus, the induced current flows in the opposite direction to the original current.

Figure 8-2C shows that the same result is found if the left-hand rule is applied to the field on the left side of the conductor. While it is true that the general direction of the field is reversed at this point, the relative motion of the conductor is also reversed so that the induced current still flows out of the page.

Figure 8-2D shows the relationship of the original current and the induced current. The original current induces a lower reverse current. The net result is that the original current is initially less than can be accounted for by Ohm's law.

The induced current is caused by an induced EMF. The induced EMF attempts to force current counter to the original current. For this reason the induced EMF is often called a *counter EMF*.

The counter EMF exists for the period of time that the magnetic field is expanding. Thus, it exists from the time that the switch is closed until the instant that the current reaches its steady state. In DC circuits, it exists only during the transient time. However, a transient condition also exists **when the switch is reopened.**

When the circuit is broken, the original current attempts to stop flowing. This causes the magnetic field to collapse. As the field collapses, it again induces an EMF into the conductor. Using the left-hand rule, we can determine the direction of the resulting induced current. Refer to Figure 8-2B again. The general direction of the magnetic field remains the same. As the field collapses inward, the conductor is cut by the flux lines as they move to the left. Thus, the relative motion is the same as if the conductor were moving to the right. Applying the left-hand rule, we find that the induced current is now in the same direction as the original current.

Of course, current cannot flow in an open circuit. Nevertheless, an EMF is induced which attempts to keep current flowing in the same direction. In some cases, the induced EMF is high enough to ionize the air between the switch contacts. In very high current circuits, the arc-over caused by the induced EMF can actually damage the switch contacts.

The process by which the induced EMF is produced is called *self-induction*. The effect of self-induction is to oppose changes in current flow. If the original current attempts to increase, self-induction opposes the increase. If the original current attempts to decrease, self-induction opposes the decrease. Self-induction may also be defined as the action of inducing an EMF into a conductor when there is a change of current in the conductor.

Inductance

Inductance is the ability of a device or circuit to oppose a change in current flow. Inductance may also be defined as the ability to induce an EMF when there is a change in current flow. *Induction* and *inductance* are easily confused, so let's discuss the difference for a moment.

Induction is the *action* of inducing an EMF when there is a change in current. Obviously then, induction exists only when a change in current occurs. Inductance is different. It is the *ability* to cause an induced voltage for a change of current. If a circuit (or device) has this ability, it has it with or without current flow. Thus, inductance is a physical property. Like resistance, inductance exists whether current is flowing or not.

The unit of measurement for inductance is the *henry* (H). It was named in honor of Joseph Henry a nineteenth century physicist who did important research in this area of science. A henry is the amount of inductance which will induce an EMF of 1 volt into a conductor when the current changes at the rate of 1 ampere per second. In most electronics applications the henry is an inconveniently large quantity. For this reason, the quantities *millihenry* (mH) and *microhenry* (μH) are more commonly used.

The symbol for inductance is L. Thus, the statement "the inductance is 10 millihenrys" can be written as an equation:

$$L = 10 \text{ mH}$$

Inductors

As we have seen, every conductor has a certain value of inductance. However, with short lengths of wire, the inductance value is so small that it can be measured only with very sensitive instruments. Many times in electronics, a specific amount of inductance is required in a circuit. A device which is designed to have a specific value of inductance is called an *inductor*.

Inductors come in various values from microhenries to several henries. The construction of the inductor is extremely simple. It consists of wire coiled around a core of some type. For this reason, the inductor is often called a coil.

Figure 8-3 shows why the inductance of a wire is increased when it is wound into a coil. In Figure 8-3A a single loop is used. As the magnetic field expands or contracts, it cuts a single turn of wire and a small value of EMF is induced. Figure 8-3B shows what happens when two loops are wound together. Notice that the field is twice as strong and that both turns are cut by the entire field. Since both the field strength and the number of turns are doubled, the induced EMF increases by a factor of 4. Thus, the inductance is four times as high. Figure 8-3C shows three turns which produce a field strength three times as high as before. Now three times the number of flux lines cut three times the number of turns. Thus, the inductance and the induced EMF increase by a factor of 3 × 3 or 9. These examples show that the inductance of the coil varies as the number of turns squared.

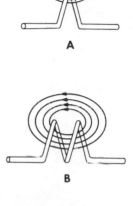

A

B

Another way to dramatically increase the inductance is to wind the coil on a core material which has a high permeability. For example, a coil wound on a **soft iron** core will have many times the inductance of an air-core coil.

Figure 8-3D shows the schematic symbol for the air-core coil or inductor. Figure 8-3E shows the symbol for the iron-core inductor. Although not indicated by the symbols, every inductor has a certain amount of resistance because the wire from which it is constructed has resistance.

C

Time Constant of an Inductor

We have seen that current cannot rise to its maximum value instantly when an inductance is in the circuit. The time required depends on the value of the inductance and the value of any series resistance. For a given value of resistance, the time required for the current to build to its maximum value is directly proportional to the value of inductance. The higher the inductance, the more time required for the current to reach maximum. On the other hand, for a given value of inductance, the time is inversely proportional to the resistance. The larger the resistance, the shorter the time required.

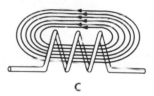

D

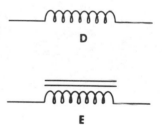

E

Figure 8-3
The Inductor

259

CAPACITANCE

Capacitance is the property of a circuit or device which enables it to store electrical energy by means of an *electrostatic* field. A device especially designed to have a certain value of capacitance is called a *capacitor*. The capacitor has the ability to store electrons and release them at a later time. The number of electrons that it can store for a given applied voltage is a measure of its *capacitance*.

Capacitors

In the early days of electronics the word *condenser* was used instead of capacitor. However, today the word condenser is rarely used except in special cases. An automobile mechanic may still call the capacitor in an ignition system a condenser but the more correct term is capacitor.

Figure 8-4 shows the principle parts of a capacitor. It consists of two metal plates separated by a non-conducting material called a *dielectric*. Often metal foil is used for the plates while the dielectric may be paper, glass, ceramic, mica or some other type of good insulator.

The actual construction of the capacitor may look quite different from that shown in Figure 8-4. For example, Figure 8-5 shows how the paper dielectric capacitor is made. Two thin sheets of metal foil are separated by a sheet of paper. Additional sheets of paper are placed on the top and bottom of the foil sheets. Then the sheets are wound into a compact cylinder. Leads are attached to each of the foil sheets. Finally, the entire unit is sealed into a permanent package by the addition of a wax or plastic case.

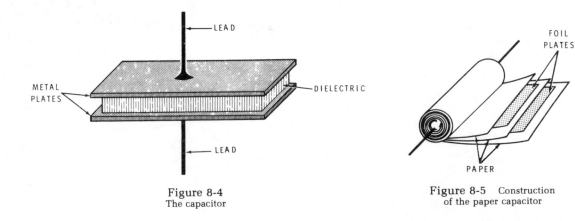

Figure 8-4
The capacitor

Figure 8-5   Construction
of the paper capacitor

Charging the Capacitor

The most useful characteristic of the capacitor is its ability to store an electrical charge. Figure 8-6 illustrates the charging action. For simplicity the capacitor is shown as two metal plates. In Figure 8-6A the capacitor is not charged. This means that there are the same number of free electrons in both plates. Naturally then, there is no difference of potential between the two plates and a voltmeter connected across the plates would read 0 volts. No current is flowing in the circuit because switch S_1 is open.

Figure 8-6B shows what happens when S_1 is closed. With S_1 closed, the positive terminal of the battery is connected to the upper plate of the capacitor. The positive charge of the battery attracts the free electrons in the upper plate. Thus, these electrons flow out of the upper plate to the positive terminal of the battery. At the same instant, the positive upper plate of the capacitor attracts the free electrons in the negative plate. However, because the two plates are separated by an insulator, no electrons can flow to the upper plate from the lower plate. Nevertheless, the attraction of the positive charge on the upper plate pulls free electrons into the lower plate. Thus, for every electron that leaves the upper plate and flows to the positive terminal of the battery another electron leaves the negative terminal and flows into the bottom plate.

As the capacitor charges, a difference of potential begins to build up across the two plates. Also, an electric field is established in the dielectric material between the plates. The capacitor continues to charge until the difference of potential between the two plates is the same as the voltage across the battery. In the example shown, current will flow until the charge on the capacitor builds up to 10 volts. Once the capacitor has the same EMF as the battery, no additional current can flow because there is no longer a difference of potential between the battery and the capacitor.

It should be emphasized again that although current flows in the circuit while the capacitor is charging, current does not flow through the capacitor. Electrons flow out of the positive plate and into the negative plate. However, electrons cannot flow through the capacitor because of the insulating dielectric. Moreover, if electrons did flow through the dielectric, the capacitor would not develop a charge in the first place. It would simply produce a voltage drop in the same way as a resistor.

Figure 8-6C shows that once the capacitor is charged, the switch can be opened and the capacitor will retain its charge. A good capacitor can retain a charge for a long period of time.

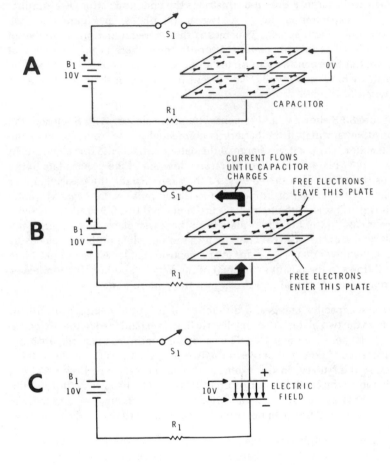

Figure 8-6
Charging the capacitor

262

Discharging the Capacitor

Theoretically, all of the energy stored in a capacitor can be recovered. Thus, a perfect capacitor would dissipate no power. It would simply store energy and later release the energy. While a perfect capacitor cannot be built, we can approach this condition. The act of storing the energy is called *charging* the capacitor. The act of recovering the energy is called *discharging the capacitor.*

Figure 8-7 illustrates the charge and discharge cycle. In Figure 8-7A, the arm of S_1 is positioned so that capacitor C_1 is placed directly across the battery. Notice the schematic symbol that is used to represent the capacitor. A current flows as shown charging the capacitor to 10 volts.

Once the capacitor is charged, let's see what happens when the arm of S_1 is moved to its other position as shown in Figure 8-7B. This removes C_1 from across the battery and places C_1 across R_2 instead. When this is done, the free electrons on the negative plate rush through R_2 to the positively charged plate. The flow of electrons continues until the two plates are once again at the same potential. At this time, the capacitor is said to be discharged and the current flow in the circuit stops.

As the capacitor discharges, the voltage across it decreases. When completely discharged the voltage across the capacitor is once again 0 volts. At this time all the energy which was initially stored has been released. The power consumed by R_2 is provided by the battery with C_1 being used as a temporary storage medium.

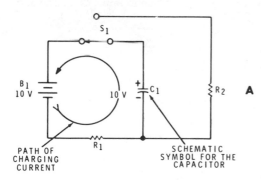 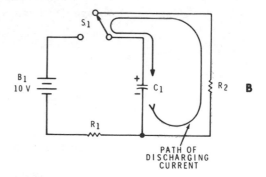

Figure 8-7
Discharging the capacitor

CAPACITORS

In many types of electronic equipment the capacitor is used more often than any other type of component except the resistor. Thus, it is important that we learn as much as possible about these devices.

Unit of Capacitance

Capacitance is a measure of the amount of charge that a capacitor can store for a given applied voltage. The unit of capacitance is the *farad* (fd). It was named in honor of Michael Faraday. One farad is the amount of capacitance that will store a charge of one coulomb when an EMF of one volt is applied. One farad is an extremely large value of capacitance. For this reason, the unit microfarad μfd or μF) meaning one millionth of a farad is more often used. Even the microfarad is frequently too large. In these cases the unit micro-microfarad ($\mu\mu$fd or $\mu\mu$F) is used. The more modern name for the micro-microfarad is the picofarad (pf or pF).

To summarize, the farad is the amount of capacitance that will store one coulomb of charge when one volt is applied. The microfarad is one millionth of a farad. The picofarad is one millionth of a microfarad or

$$\frac{1}{1,000,000,000,000}$$

of a farad. Using powers of ten, the microfarad is 10^{-6} farad while the picofarad is 10^{-12} farad.

There is a formula which expresses capacitance in terms of charge and voltage. The formula is:

$$C = \frac{Q}{E}$$

C is the capacitance in farads; Q is the charge in coulombs; and E is the EMF in volts.

Factors Determining Capacitance

Capacitance is determined by three factors:

1. the area of the metal plates,
2. the spacing between the plates, and
3. the nature of the dielectric.

264

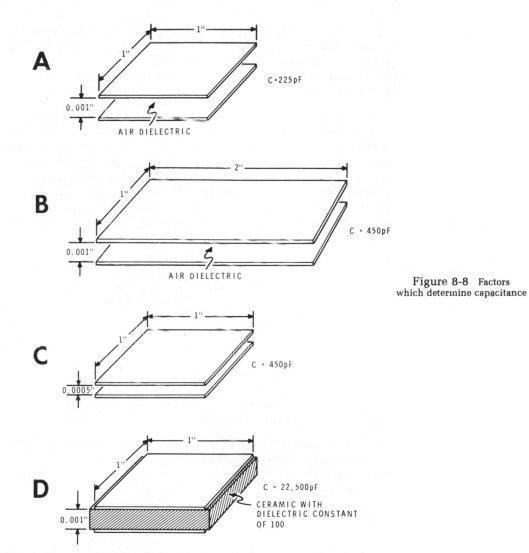

Figure 8-8 Factors which determine capacitance

This is illustrated in Figure 8-8. Figure 8-8A shows a capacitor made up of two 1-inch square plates separated by 0.001 inch. The dielectric is air. Such a device has a capacitance of 225 pf. Let's use this capacitor as a reference and see what happens when we change the area of the plates, the spacing of the plates, and the nature of the dielectric.

Figure 8-8B shows what happens when the area of the plates is doubled but all other factors are held constant. Notice that there is now twice as much area in which the electrostatic field can exist. This doubles the capacity or capacitance of the device. The capacitance value doubles to 450 pf. Thus, capacitance is directly proportional to the area of the plates.

265

Figure 8-8C shows that the capacitance can also be doubled by reducing the spacing between the plates to one half its former value. The path of the electrostatic lines of force is cut in half. This doubles the strength of the field which, in turn, doubles the capacitance. Thus, capacitance is inversely proportional to the spacing between the plates.

Finally, Figure 8-8D shows that the capacitance can be greatly increased by using a better dielectric than air. Air makes a very poor dielectric. Most insulators support electrostatic lines of force more easily than air. The ease with which an insulator supports electrostatic lines of force is indicated by its *dielectric constant*. Air is used as a reference. It is arbitrarily given a dielectric constant of 1. Most insulators have a higher dielectric constant. For example, a sheet of waxed paper has a dielectric constant of about 3. This means that a sheet of 0.001 inch waxed paper placed between the plates would triple the capacitance.

Some typical dielectric constants for common types of insulators are:

Material	Dielectric Constant (K)
Air	1
Vacuum	1
Waxed Paper	3 — 4
Mica	5 — 7
Glass	4 — 10
Rubber	2 — 3
Ceramics	10 — 5000

Figure 8-8D shows that the capacitance value can be increased by a factor of 100 simply by inserting between the plates an insulator with a dielectric constant of 100. Thus, capacitance is directly proportional to the dielectric constant.

There is a formula which combines the three factors discussed above. It is:

$$C = 0.225 \ K \ \frac{A}{d}$$

C is the capacitance in picofarads; K is the dielectric constant; A is the area (in square inches) of one plate; and d is the distance (in inches) between the two plates.

Types of Capacitors

Capacitors are available in many different shapes and sizes. However, all capacitors can be placed in one of two categories: variable and fixed. Let's discuss these two categories in more detail.

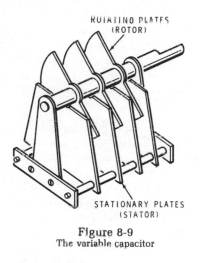

ROTATING PLATES
(ROTOR)

STATIONARY PLATES
(STATOR)

Figure 8-9
The variable capacitor

Variable Capacitors. Figure 8-9 shows the construction of an air-dielectric variable capacitor. The capacitance value of this type of capacitor can be changed by rotating the shaft. The rotating plates are attached to the shaft. As the shaft is turned, the rotating plates change position in relation to the stationary plates. The rotating plates are electrically connected and form one plate of the capacitor. The stationary plates are also electrically connected. They form the other plate of the capacitor. The rotating plates and stationary plates mesh together but do not touch. By moving the shaft, the area of the plates across from each other can be changed from maximum when fully meshed to minimum when fully open. As we have seen, this changes the capacitance of the device.

Fixed Capacitors. Most fixed capacitors are constructed as shown earlier in Figure 8-5. They consist of alternate layers of metal foil (plates) and insulators (dielectric).

Capacitors are often named for their dielectric. Thus, there are paper, ceramic, and mica capacitors. Also, capacitors are sometimes classified according to their shape. Thus, there are disc capacitors and tubular capacitors.

267

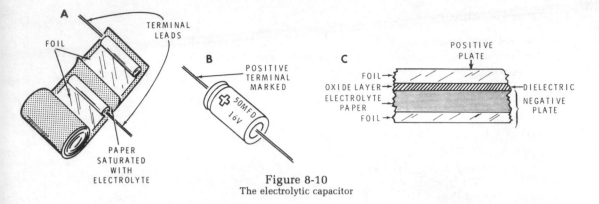

Figure 8-10
The electrolytic capacitor

One of the most popular types of fixed capacitor is the electrolytic capacitor. Its construction is illustrated in Figure 8-10. Sheets of metal foil are separated by a sheet of paper or gauze which is saturated with a chemical paste called an electrolyte. The electrolyte is a good conductor and therefore the paper is not the dielectric. Actually, the dielectric is formed during the manufacturing process. A DC voltage is applied across the foil plates. As current flows, a thin layer of aluminum oxide builds up on the plate which is connected to the positive side of the DC voltage. As shown in Figure 8-10C, the oxide layer is extremely thin. Because the oxide is a fairly good insulator it acts as a dielectric. The upper foil becomes the positive plate; the oxide becomes the dielectric; and the electrolyte becomes the negative plate. Notice that the bottom layer of foil simply provides a connection to the electrolyte.

Recall that capacitance is inversely porportional to the spacing between plates. Because the oxide layer is extremely thin, very high values of capacitance are possible with this technique. While most other capacitors have values below 1 μfd, the electrolytic capacitor may have values ranging from about 5 μfd up to thousands of microfarads.

Because of its construction, the electrolytic capacitor is *polarized*. This means that the capacitor has a negative and a positive lead. When connected in a circuit, the positive lead must be connected to the more positive point. As shown in Figure 8-10B, the positive lead is marked on the capacitor.

An important characteristic of the electrolytic (or any other type of capacitor) is its voltage rating. The voltage rating indicates the maximum voltage that the capacitor can withstand without the dielectric breaking down or arcing over. With electrolytic capacitors this value is generally printed on the capacitor along with its capacitance value.

RC TIME CONSTANTS

When a capacitor is connected across a DC voltage source, it charges to the applied voltage. If the charged capacitor is then connected across a load, it will discharge through the load. The length of time required for a capacitor to charge or discharge can be computed if certain circuit values are known.

There are only two factors which determine the charge or discharge time. These are the value of the capacitor and the value of the resistance through which the capacitor must charge or discharge. The time is directly proportional to both resistance and capacitance. To understand the relationship between resistance, capacitance, and time, we must consider the idea of an RC time constant.

A time constant is the time required for a capacitor to charge to 63.2 percent of the applied voltage. Or if the capacitor is being discharged, a time constant is the length of time required for the voltage across the capacitor to drop by 63.2 percent.

The time constant can be expressed as an equation:

$$t = R \times C$$

Here t is the time constant in seconds (time required to reach 63.2 percent of full charge); R is the resistance in ohms; and C is the capacitance in farads. As mentioned earlier the farad is too large to be practical and capacitance is most often expressed in microfarads. In the above equation, if C is in microfarads and R is in ohms, then t will be in microseconds. If C is in microfarads and R is in kilohms, then t will be in milliseconds. Finally, if C is in microfarads and R is in megohms, then t will be in seconds.

Some examples may help illustrate this.
If C = 1 μfd and R = 100Ω; then:
    t = R × C
    t = 100Ω × 1 μfd
    t = 100 microseconds

If C = 1 μfd and R = 10 kΩ; then:
    t = R × C
    t = 10 kΩ × 1 μfd
    t = 10 milliseconds

If C = 1 μfd and R = 2 MΩ; then:
    t = R × C
    t = 2 MΩ × 1 μfd
    t = 2 seconds

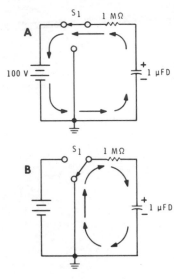

As you study these examples keep in mind that the time constant (t) is *not* the time required to fully charge (or discharge) the capacitor. Rather, it is the time required to charge the capacitor to 63.2 percent of the applied voltage. To see how a capacitor charges, let's consider a specific example. Figure 8-11 shows a 1 μfd capacitor connected in series with a 1 MΩ resistor. Thus, the time constant is 1 second. Initially, the capacitor is completely discharged and the voltage across it is 0 volts. Now let's see how the capacitor charges when the arm of the switch is moved up so that the 100-volt power supply is connected to the R-C network.

When the capacitor is connected across the voltage source, it attempts to charge to the level of the applied voltage. However, the capacitor does not charge instantaneously. The time constant of the circuit is:

$$t = R \times C$$
$$t = 1\ M\Omega \times 1\ \mu fd$$
$$t = 1\ second$$

Thus, after 1 second (1 time constant), the capacitor will have charged to 63.2 percent of the applied voltage or to 63.2 volts.

Figure 8-11
Working with time constants

Figure 8-12 shows two curves which are helpful when working with time constants. Curve A shows how a capacitor charges. Initially, the capacitor charges rapidly, charging to 63.2 percent of the applied voltage during the first time constant. However, as time passes, the capacitor begins to charge more slowly.

During the second time constant the capacitor charges to 63.2 percent of the remaining voltage. In our example, the remaining voltage after 1 time constant (1 second) is 100 V − 63.2 V = 36.8 V. Now, 63.2 percent of 36.8 V is about 23.3 V. Thus, at the end of the second time constant (after 2 seconds) the voltage on the capacitor has risen to: 63.2V + 23.3V = 86.5 V. This is 86.5 percent of the applied voltage.

During the third time constant, the capacitor once again charges to 63.2 percent of the remaining voltage. After 3 time constants (3 seconds), the capacitor has charged to 95 percent. After 4 time constants (4 seconds), it has charged to 98.2 percent; and after 5 time constants to more than 99 percent. For most purposes, the capacitor is considered fully charged after five time constants.

Because of its shape, curve A is called an exponential curve. The capacitor is said to charge exponentially.

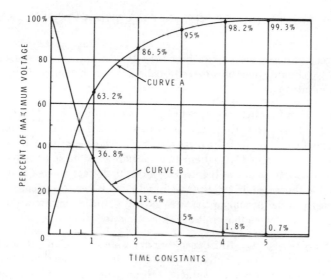

Figure 8-12
Time constant curves

Curve B of Figure 8-12 shows how the capacitor discharges. At the first instant the capacitor is fully charged. During the first time constant, the voltage drops by 63.2 percent to 36.8 percent of its original value. During the second time constant, the voltage drops an additional 63.2 percent or to only 13.5 percent of its original value. The voltage drops to about 5 percent at three time constants and to only 1.8 percent after four. After five time constants, the charge is less than 1 percent of its original value. For most practical purposes, the capacitor can be considered fully discharged after five time constants.

In Figure 8-11A the capacitor is charged to approximately 100 volts in 5 seconds. Figure 8-11B shows the capacitor being discharged. According to curve B, the charge on the capacitor will be:

    100 volts initially;
    36.8 volts after 1 time constant (1 second);
    13.5 volts after 2 time constants (2 seconds);
    5 volts after 3 time constants (3 seconds);
    1.8 volts after 4 time constants (4 seconds);
    and 0.7 volts after 5 time constants (5 seconds).

271

CAPACITORS IN COMBINATION

Like resistors, capacitors can be connected in various combinations. Thus, we should study how capacitors behave when connected together in different ways.

Capacitors in Parallel. Figure 8-13A shows the dimensions of the 225 pf capacitor discussed earlier. In Figure 8-14B, two of these capacitors are connected in parallel. Of the three factors which determine capacitance, only one has changed. The dielectric constant and the spacing between the plates are the same as before. However, the effective area of the two plates has increased. In fact the area has doubled. Recall that capacitance is directly proportional to the area of the plates. Therefore, the total capacitance is twice that of the single capacitor.

This shows that connecting capacitors in parallel is equivalent to adding the plate areas. Consequently, the total capacitance is equal to the sum of the individual capacitance values. If three capacitors are connected in parallel, the total capacitance (C_t) is found by adding the individual values:

$$C_t = C_1 + C_2 + C_3$$

If $C_1 = 5$ μfd, $C_2 = 10$ μfd, and $C_3 = 1$ μfd; then $C_t = 16$ μfd. Notice that capacitors in *parallel* add like resistors in *series*.

Capacitors connected in parallel will all charge to the same voltage. Remember the voltage is the same across every section of a parallel network.

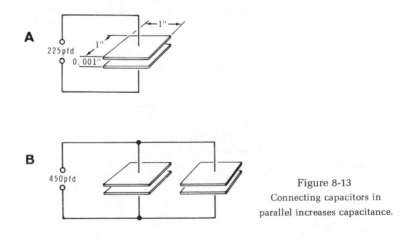

A
225pfd
1"
1"
0.001"

B
450pfd

Figure 8-13
Connecting capacitors in
parallel increases capacitance.

272

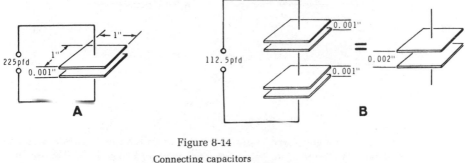

Figure 8-14
Connecting capacitors
in series decreases capacitance.

Capacitors In Series. Figure 8-14 compares our "standard" capacitor with two "standard" capacitors connected in series. As shown, this is equivalent to doubling the thickness of the dielectric. That is, two series capacitors act like a single capacitor which has a dielectric thickness equal to the sum of the individual dielectric thicknesses. Since capacitance is inversely proportional to the spacing between the plates, doubling the thickness of the dielectric cuts the total capacitance value to one half that of the "standard" capacitor.

The total capacitance of a group of series capacitors is calculated in the same way as the total resistance of parallel resistors. Or stated more simply; capacitors in series combine in the same way as resistors in parallel. The total capacitance of two capacitors in series can be calculated by the formula:

$$C_t = \frac{C_1 \times C_2}{C_1 + C_2}$$

Notice that this equation has the same form as the equation for calculating two resistors in parallel.

When more than two capacitors are connected in series, the following formula is used:

$$C_t = \frac{1}{\dfrac{1}{C_1} + \dfrac{1}{C_2} + \dfrac{1}{C_3} + \dots}$$

Here again, the equation has the same form as the one shown earlier for calculating total resistance in parallel circuits.

273

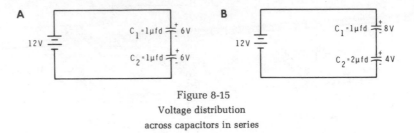

Figure 8-15
Voltage distribution
across capacitors in series

Figure 8-15 shows two capacitors connected in series across a 12-volt battery. An interesting thing about capacitors in series is the way in which the applied voltage is distributed across the capacitors. If the two capacitors have the same value, then the applied voltage is distributed evenly between the two. In Figure 8-15A, each capacitor charges to one half the applied voltage or to 6 volts.

However, when the capacitors have different values an interesting thing happens. The smaller capacitor charges to a higher voltage than the larger capacitor. In Figure 8-15B, C_2 is twice as large as C_1. Therefore C_1 charges to twice the voltage. Since the total voltage across both capacitors must be 12 volts, C_1 must drop 8 volts while C_2 drops only 4 volts.

To understand why the voltage is distributed in this way, we must recall an earlier equation which expressed capacitance in terms of charge and voltage. The equation is:

$$C = \frac{Q}{E}$$

Where C is capacitance in farads, Q is the charge in coulombs, and E is in volts. This equation can be rearranged:

$$E = \frac{Q}{C}$$

In this form, the equation states that the voltage across the capacitor is directly proportional to the charge on the capacitor but inversely proportional to the size of the capacitor. In Figure 8-15B, the two capacitors are in series. Consequently, the charging current is the same at all points in the circuit. For this reason, the two capacitors must always have equal charges. Since the charges are equal, the voltage is determined solely by the value of the capacitor. And since the voltage is inversely proportional to the capacitance value, the smaller capacitor will charge to a higher voltage.

SUMMARY

When current flows through a conductor a magnetic field builds up around the conductor. As the magnetic field builds up, a voltage is induced into the conductor. The induced voltage opposes the applied voltage and is called a counter EMF. The process by which the counter EMF is produced is called self-induction.

The counter EMF is always of a polarity which opposes changes in current. It opposes the increase in current which occurs when power is applied to a circuit. It also opposes the decrease in current which occurs when power is removed. The ability of a device to oppose a change of current is called inductance. The unit of inductance is the Henry.

A device designed to have a specific inductance is called an inductor. It consists of turns of wire wrapped around a core. The greater the number of turns and the higher the permeability of the core, the greater the value of inductance will be.

The capacitor consists of two metal plates separated by an insulator called a dielectric. It has the ability to store an electrical charge. This ability to store a charge is called capacitance. When connected to a voltage source, the capacitor will charge to the value of the applied voltage. If the charged capacitor is then connected across a load, it will discharge through the load.

The unit of capacitance is the farad although the microfarad (μfd or μF) and the picofarad (pf or pF) are more commonly used. Three factors determine the value of a capacitor. They are: the area of the plates, the spacing between the plates, and the dielectric constant.

There are many different types of capacitors. They are generally classified by their dielectric. The most popular types are air, paper, mica, ceramic, and electrolytic.

The length of time required for a capacitor to charge is determined by the capacitance and the resistance in the circuit. A time constant is the length of time required for a capacitor to charge to 63.2 percent of its applied voltage. The formula for finding the time constant is:

$$t = R \times C$$

The time constant chart is used when working with time constants. It shows the manner in which capacitors charge and discharge. It plots the number of time constants against the percent of the applied voltage.

Capacitors may be connected in series or in parallel. When connected in parallel, the total capacitance is equal to the sum of the individual capacitor values. The formula is:

$$C_t = C_1 + C_2 + C_3 + \ldots$$

When connected in series, the total capacitance is determined by the following formula:

$$C_t = \cfrac{1}{\cfrac{1}{C_1} + \cfrac{1}{C_2} + \cfrac{1}{C_3} + \ldots}$$